The Sirt Food Diet

Eat your Way to Rapid Weight Loss and a Longer Life by Triggering the Metabolic Power of Skinny Gene. Includes Quick and Healthy Recipes for your Meal Plan 2020

ADELE ADKINS

© Copyright 2020
All rights reserved

TABLE OF CONTENTS

INTRODUCTION	9
CHAPTER ONE	11
WHAT IS THE SKINNY GENE	11
What Is the Skinny Gene-Diet?	11
Why Is It Called the Sirt Diet?	12
How Does the Sirt Diet Work?	13
The Allied Foods of Our Metabolism	14
The Benefits	15
This Is the Right Way to Get Started	16
Is It Effective?	16
CHAPTER TWO	19
THE PHASES OF THE SIRTFOOD DIET	19
First Phase	19
Second Phase	20
Contraindications	21
Disadvantages of the Sirtfood Diet	21
Are "Sirtfood" diets scientifically sound?	22
Diet Under the Microscope	23
The Science of Weight Loss	24
Is there a Quick Fix?	26
CHAPTER THREE	29
Sirtuins and Weight Loss	29
The Most Important Sirt Foods for Losing Weight	30
Physical Activity and the Sirt Food Diet	31
Foods that Activate the Lean Gene	32

Foods that Increase Physical Performance 36

7-Day Meal Plan 37

CHAPTER FOUR 41

SIRT FOOD DIET: MYTH OR TRUTH 41

Who Came Up with this? 41

How Does the Sirtuin Diet Work? 42

What Is the Catch? 42

Are Sirt-Foods New Superfoods? 42

Is the Sirtuin Diet Healthy? 43

Does the Sirtuin Diet Work? 44

Do You Need to Take Dietary Supplements on a Sirt Food Diet? 44

When It Is Appropriate to Do it 45

When there Is No Need to Do It 45

What to Drink on a Sirt Food Diet? 45

CHAPTER FIVE 47

SIRT FOOD BREAKFAST RECIPES 47

    Honey Cake with Orange Cream 47

    Banana Bread with Walnuts 49

    Currant and Banana Croissants with Ground Almonds 50

    Pea Protein Sandwiches 52

    Sweet Pumpkin Buns 54

    Avocado Smoothie with Yogurt and Wasabi 55

    Goat Cheese Omelet with Arugula and Tomatoes 56

    Baked Quark Toasts with Orange Fillets 58

    Sweet Millet Casserole with Clementine 59

    Cloud Bread 60

    Cottage Cheese with Raspberry Sauce 62

    Oriental Porridge with Oranges and Figs 63

Blackberry and Apple Spread with Lavender Flowers 64

Hearty Smoked Salmon Slices with Cream Cheese and Onion 66

CHAPTER SIX 67

Sirt Food Lunch Recipes 67

Vegan Raw Meatballs 67

Keto Bowl with Mushrooms and Chinese Cabbage 68

Spaghetti with Pumpkin and Spinach and Goat's Cream Cheese Sauce 70

Salad Boat with Chickpeas and Tzatziki 71

Stuffed Peppers with Quinoa, Ricotta and Herbs 73

Salmon with Herb and Walnut Salsa 74

Jackfruit Fricassee with Pea Rice 75

Dill Patties with a Dandelion Dip 77

Konjac Pasta with Berries 79

Carrot Tagliatelle with Avocado Pesto 80

Strawberry and Avocado Salad with Chicken Nuggets 82

Sweet Potatoes with Asparagus, Eggplant and Halloumi 84

CHAPTER SEVEN 87

SIRT FOOD DINNER RECIPES 87

Pulled Chicken in Salad Tacos 87

Veal cabbage Rolls – Smarter with Capers, Garlic and Caraway Seeds 88

Fish Cakes with Potato Salad 90

Egg and Avocado Sandwich with Crab Salad 91

Vegetarian Lasagna - Smarter with Seitan and Spinach 93

Spaghetti with Salmon in Lemon Sauce 95

Rice Shakshuka with Olives 96

Low Carb Pancakes with Spinach and Cheese 98

Baru Nut Bowl 99

Grilled Rosemary Sea Bream 101

Poached Eggs on Spinach with Red Wine Sauce 103

Pasta with Minced Lamb Balls and Eggplant, Tomatoes and Sultanas 104

CHAPTER EIGHT 107

DESSERT RECIPES 107

Blackberry Quark Tartlets 107

Chocolate Strawberries with Cardamom 108

Chocolate Fruit Cake 109

Pear Chocolate Cake with Pistachios 110

Nut and Chocolate Cookies 112

Cauliflower and Cheese Gressinos 113

Apple Cheesecake 114

Halloween Desserts 115

Zucchini and Cheese Quiche 117

Asparagus Quiche 118

Spelled Waffles with Cherry Sauce 119

Strawberry Cream Cake 121

Brownie Cheesecake 123

CHAPTER NINE 125

SNACKS RECIPES 125

Pancake Skewers with Fruits 125

Tomato and Zucchini Salad with Feta 126

Blueberry and Coconut Rolls 127

Brain Food Cookies 128

Chocolate Granola Bars 129

Blueberry and Coconut Rolls 131

Vanilla Energy Balls with Coconut Shell 132

Wake-Up Energy Balls 133

Brain Food Cookies 134

Chocolate Granola Bars 135

Kale Avocado and Chili Dip with Keto Crackers 136

Grilled Eggplant Rolls with Walnut and Feta Filling 137

CHAPTER TEN 141

Fruit Juice 141

Drinks with Oranges 141

Orange and Mandarin Liqueur 141

Pear and Lime Marmalade 142

Kiwi Yogurt Ice Cream 143

Christmas Cocktail - Vegan Eggnog 144

Watercress Smoothie 145

Cucumber and Orange Drink 146

Green Smoothies with Yogurt 147

CHAPTER ELEVEN 149

SALAD AND SOUP RECIPES 149

Cauliflower Soup with a Mackerel Fillet 149

Arugula Cream Soup with Parmesan 150

Cold Melon and Tomato Soup with Yogurt and Basil 152

Tomato Soup with Roasted Buckwheat 153

Piche Steiner Stew 154

Chicken Soup the Grandmother's Way 156

Caldo Verde - Portuguese Kale Soup 158

Avocado and Mozzarella Salad Bowl 159

Chicory and Orange Salad 160

Mango and Avocado Salad with Watercress 160

Coconut Pancakes with Kiwi Salad 162

Cucumber and Radish Salad with Feta 163

May Beet Salad with Cucumber 164

Lentil Salad with Spinach, Rhubarb and Asparagus 165
May Beet Salad with Cucumber 166
CONCLUSION 168

INTRODUCTION

The Sirt Diet is going through its hype, as shown by Google statistics. It owes its fame to the spectacular transformation of celebrities. Recently, the Sirt Diet has even been called the "Adele diet." Another reason it is so popular maybe is that you can eat chocolate and drink wine on the Sirtuin Diet.

According to the authors of the diet, caloric restrictions and a recommended group of products are designed to extend the life and "accelerate metabolism."

Sirtuins or Sir proteins (Silent information regulator) are enzymes responsible for the cleavage reactions of acetyl groups from different classes of proteins. Recently, they have aroused great interest. Studies have shown that reducing dietary intake by 30-50% of calories resulted in longer life in yeast by activating the Sir protein.

Scientists immediately began to wonder how sirtuins affected the human body. It turns out they can act as factors regulating the rate of aging and affect DNA repair processes. Accordingly, they began to be called the "elixir of youth."

The Sirt Food Diet is a diet that is supposed to ensure long life and a slim figure. The menu is dominated mainly by-products rich in polyphenols activating sirtuins - so-called longevity proteins, and green cocktails prepared on their basis. According to the authors, the effects of the Sirt

Diet in slimming are impressive (about 3 kg a week). Is it really worth using such a diet? You find the benefits in this book.

CHAPTER ONE

What Is the Skinny Gene

This new diet is coming directly from Great Britain and Ireland, where it is gathering more and more approval, even among celebrities! This is the Sirt Diet, the so-called lean gene-diet, which without particular products, but only thanks to the intake of certain foods, allows us to lose weight and lose a few extra pounds in a short time.

The SIRT Diet, also known as the "lean gene-diet," has caused singer Adele to lose 30 kilos in a year: now she is relocating on the Net, and for this reason, the Centro San Camillo of Bari has decided to clarify this topic.

The SIRT Diet, explains Dr. Marika De Tullio of our Nutrition Biology clinic, is the diet of the moment. According to its creators, nutritionists Aidan Goggins and Glen Matten, it would allow you to lose almost three and a half pounds in the first week and then maintain your ideal weight forever.

What Is the Skinny Gene-Diet?

The Sirt Diet was born in the UK thanks to two nutritionists, Aidan Goggins and Glen Matten, discoverers of particular benefits of some foods capable of activating sirtuins, a group of genes that can speed up the metabolism, making you burn fat and promoting quick weight loss: there is talk of 3 kilos lost during the first week!

Furthermore, compared to other food regimes, it includes foods, it does not exclude them: this means that even if we have to commit some sacrifices, all the different nutrients that our body needs never go missing.

The authors of the Sirt Food Diet recommend that it should be used by people whose BMI is greater than 20. The Sirt Food Diet aims to achieve weight reduction, which is why it is addressed mainly to people who are overweight and obese. If you suffer from chronic diseases, you should consult a doctor before starting it.

Due to the fact the first phase is low in calories, the use of the Sirt Food Diet by people with very low body weight can be dangerous to their health. The Sirt Food Diet is not recommended for children, women planning conception, or pregnant and lactating women because it may turn out to be deficient in some ingredients - in particular, Phase 1 of the Sirt Food Diet. However, we can include sirtuin products on the menu because they have a very beneficial effect on health.

Why Is It Called the Sirt Diet?

This weight loss diet is called SIRT because it is based on the action of a group of proteins called sirtuins. Our cells build seven different sirtuins (SIRT1 - SIRT7). They perform a variety of functions, including that of regulating the transcription of genes, that of binding to proteins in the cytoplasm and mitochondria, and the function of regulating a large variety of processes related to metabolism, stimulating it more and related to neurodegeneration.

In many organisms, they are implicated in the phenomenon of aging. This discovery has triggered an intense search for compounds that can somehow increase the enzymatic activity of the sirtuins themselves, such as, e.g., resveratrol, a natural polyphenol found in some fruits, but especially in grape skins.

How Does the Sirt Diet Work?

The Sirt Diet is mainly divided into two phases: the first lasts a week and the other 14 days.

Phase 1 is the most intense, where you see the most results and it allows you to lose up to 3.2 pounds. The maximum of calories that can be consumed during the first three days is 1,000, while from the fourth to the seventh, one reaches 1,500 calories per day.

The menu to follow includes a "fixed" part, the one relating to green juice designed by nutritionists that helps to moderate the appetite of the brain, and one that varies daily. The green juice recipe is simple and includes all-natural products: 75 g of curly kale, 30 g of arugula, and 5 g of parsley must be blended together with 150 g of green celery with the leaves and 1/2 green apple, grated. Everything must be completed with half a squeezed lemon and half a teaspoon of matcha green tea.

Monday-Wednesday: 3 Sirt green juices to be taken on waking up, mid-morning and mid-afternoon; 1 solid meal of animal or vegan protein (for example, turkey escalope or buckwheat noodles with tofu) accompanied by vegetables, always ending with 15-20 g of 85% dark chocolate.

Thursday-Sunday: 2 Sirt green juices and two solid meals, remembering to always vary the main course chosen, from salmon fillet to vegetable tabbouleh to buckwheat spaghetti with celery and kale.

Phase 2 is used to maintain weight. During this period, the goal is the consolidation of weight loss, although the possibility of losing weight is not excluded. To do all this, just eat the exceptional foods rich in sirtuins.

The Allied Foods of Our Metabolism

If sirtuins, the genes of thinness, are found in certain foods, it will mean that we will lose weight just by eating! Our best friends to achieve this are:

- Strawberries
- Blueberries
- Citrus fruits
- Cabbage
- Rocket salad
- Mountain celery (also called lovage)
- Parsley
- Red onion

- Tofu
- 85% dark chocolate
- Chili pepper
- Turmeric
- Extra virgin olive oil

Without forgetting matcha green tea, already our ally in many other daily challenges, passion fruit, walnuts, red wine, radicchio and coffee. In short, every day, there will be colors, variety and taste!

The Benefits

Compared to many other unbalanced diets, which focus more on weight loss, the Sirt Diet plans to take foods that excellently regulate the entire metabolism and burn fat, in addition to improving the health of the cells and promoting the increase of muscle mass, which turns out to be perfect for those who already train constantly in the gym.

Furthermore, we go to act "from the inside," given that the role of sirtuins consists above all in checking that the genes remain active or inactive in particular situations. Hence, when DNA damage occurs, they not only regulate the repair mechanism, but improve the general health of the cells. For this reason, more and more studies are being carried out on these allies of ours, since they seem to be markers of longevity! However, remember that this diet should not always be followed: for some people, it will be necessary to repeat it only twice a year, while for others once every three months. This depends on the build, the metabolism and many factors that vary in each of us. Our advice is always to go to a specialist when deciding to start a new diet. Health comes first!

This Is the Right Way to Get Started

The first phase of the Sirt Food Diet, with a calorie intake of 1,000 kilocalories, is probably particularly difficult for many people. To make getting started easier, you can also make a slow calorie reduction. This means that the usual number of calories is not immediately reduced to 1,000 calories, but is slowly adjusted depending on the physical well-being (but should not extend over a long period of time).

A gentler change makes it easier to hold out, and the body can gradually get used to the new diet. However, weight loss is somewhat delayed.

It is also important to inform yourself well beforehand about the Sirt Food Diet and to get an overview of sirtuin foods. So, you can plan your meals precisely without getting stressed.

Is It Effective?

The Sirtfood Diet's writers boldly say the diet will overwhelm weight loss, trigger the "lean gene," and prevent illness.

The thing is there is not much evidence to back it up.

There is thus far no convincing evidence that the Sirtfood Diet has a more beneficial effect on weight loss than any other low-calorie diet.

And while many of those foods have health benefits, no long-term human studies have been conducted to determine whether eating a diet rich in sirt foods has any measurable health benefits.

Sirtfood Diet nevertheless presents the results of a pilot study carried out by the authors and involving 39 participants from their fitness center. The results of this study do not seem, however, to have been published elsewhere.

Participants followed the diet for a week and were exercising daily. Participants lost an average of 3.2 kg (7 pounds) at the end of the week and retained or even added muscle mass.

These tests, however, are hardly shocking. Limiting your calorie intake to 1,000 calories and actively exercising will almost always lead to a weight loss.

Either way, this form of rapid weight loss is neither genuine nor permanent, and after the first week, this study did not follow the participants to see if they had gained weight, which is usually the case.

As well as consuming fat and muscle, when the body is hungry for nutrition, it uses its emergency energy stores of glycogen.

Each glycogen molecule needs to store 3-4 molecules of water. It also gets rid of the water as the body uses glycogen. This is called "water weight."

Only about one-third of the weight loss comes from fat during the first week of intense calorie restriction, while the other two-thirds come from the skin, muscle, and glycogen.

The body replenishes its glycogen stores as soon as the calorie intake increases, and the weight returns immediately.

Unfortunately, this form of calorie restriction can also lower the metabolic rate of your body, so you need even fewer energy calories per day than before.

This diet is likely to help you lose a few pounds at first, but it'll probably come back as soon as the diet is over.

When it comes to the prevention of disease, three weeks is probably not long enough to have a long-term measurable impact.

On the other hand, it may well be a good idea to add Sirt foods to your regular long-term diet. Yet miss the diet in this situation, and start doing so now.

CHAPTER TWO

The Phases of the Sirtfood Diet

There is always, however, a dark point for any diet that is proposed to be miraculous and easy to follow: it is highly low-calorie in the first phase or initial phase, which lasts seven days and which involves the loss of about 3 kg.

First Phase

In fact, according to the creators of the diet, the first phase of the diet involves a drastic reduction in calories: in the first three days, Goggins and Matten recommended to stay below 1,000 calories (very few), eating three green juices in the morning, at mid-morning and mid-afternoon, and eating one solid meal.

Blending or passing the food through an extractor, the authors of the diet claim makes the active ingredients of the SIRT plants more absorbable. So it is possible to blend, according to a proposed recipe, two abundant handfuls of curly cabbage, parsley and arugula, and then, to blend ½ green apple with 2-3 stalks of green celery and to add in a ½ teaspoon of matcha tea powder.

In the following four days, the regime goes to about 1,500 calories with the addition of a second solid meal represented by buckwheat pasta, prawns or chicken breast with curly cabbage and red onions, etc. Every day it is necessary to eat a main SIRT food in this diet, which is 15-20 g of 85% dark chocolate.

The consequence of this drastic calorie reduction can only be a loss of about 3 kg without stressing yourself with intense or prolonged sports activities and without too many sacrifices. But, in reality, at least in the initial phase of the diet, sacrifices and disadvantages are there: the mere fact of drinking blended fruit and vegetables with only 1-2 solid meals a day is not easy. Healthy foods are certainly those proposed to be extracted and blended, but doing so risks losing the pleasure of food and being at the table, but also ruins a good mood.

The consolation, say the creators of the diet, is that the sense of hunger is not felt, because SIRT foods act on the idea of satiety by suppressing hunger. But it is anything but simple because at the beginning hunger is the first sensation that is perceived in a completely natural and physiological way: going from a dietary regimen of, e.g., 2,000 kcal to one with less than 1,000 kcal necessarily involves a response from the body which translates into a feeling of hunger.

Second Phase

The second phase of the diet lasts two weeks and is aimed at consolidating the weight loss of the previous period (it is the so-called maintenance phase), where three meals a day, one green juice, and two snacks are consumed. There are no more restrictions on calories, but only further recommendations to take other SIRT foods, namely cabbage, dark chocolate, red wine, citrus fruits, coffee, blueberries, capers, green tea, soy, strawberries, eggs, yogurt, chicken, prawns, beans, potatoes, pasta, pizza, and SIRT salads, as well as fruit and vegetables. Two or three times a week, with meals, you can drink a glass of red wine: it contains polyphenols, such as resveratrol, which, as mentioned, activates sirtuins.

Contraindications

The Sirt Diet aims for rapid weight loss. However, many experts disagree on this food plan. According to some, this rapid weight loss can have a long-term counterproductive effect: slower metabolism and sudden weight gain. Many experts also say that following this eating plan for a long time can lead to emotional repercussions, such as excessive irritability.

Above all, it is essential to consult your doctor and count on a nutritionist to establish the right diet plan for you. This is because each person is different; there are those who suffer from certain pathologies or even have mild ailments.

The advice of a specialist is the first action to take, especially in case of pregnancy or breastfeeding.

Disadvantages of the Sirtfood Diet

The Sirtfood Diet reduces your daily calorie intake to just 1,000 during the "first phase," which can cause fainting, anemia or even lower your blood pressure. The risks of affecting your health accumulate when you mix food and sports routines.

Nor is it a long-term eating plan, because the Sirtfood Plan is designed to get results in a short time and is not focused on improving your eating habits.

As promising as this new miracle diet sounds, if you take it seriously, you will lose weight with every diet if only 1,500 kcal per day is allowed. The sirtuins will undoubtedly support weight loss and thus help you lose weight, but in our opinion, they are not decisive for a weight loss that should be three kilograms per week.

And even though Sirtfood is basically not about counting calories, you have no other option if you want to adhere to the specified calorie limits stringently. Since the limits are quite strict, the yo-yo effect threatens your loss after the end of the diet, so you should be careful. Incidentally, meat is in pointless in the Sirtfood Diet - a tough lot for meat lovers!

Are "Sirtfood" diets scientifically sound?

Nutrition is an atmosphere where a lot of irrelevant or misleading advice is gravitating, from needless items to eccentric and unproven claims to simple scams. Any new diet that comes onto the market, such as a trend, must, therefore, be viewed with caution, without relying on testimonies that are either fabricated or that only represent a personal opinion that do not prove anything. The new trendy diet that makes headlines is the "SirtFood" diet, which will help you to lose weight while bringing other benefits like "stimulating rejuvenation and cell reparation!"

For the uninitiated, this diet is based on eating foods that might interact with a family of proteins known as sirtuin proteins, or SIRT1-SIRT7. This diet is all the more appealing as it includes authorized foods such as wine and chocolate, which are supposed to be the best sources, such as citrus fruits, blueberries or even cod.

The calorie intake is limited during the first three days (1,000 calories per day) and consists of three vegetable juices called "SirtFood," plus a normal meal rich in "Sirtfoods." Calorie intake rises to 1,500 calories from the fourth day to the seventh day, which consists of two drinks and two meals. Beyond that, a healthy diet rich in sirtuin foods, along with vegetable juices, is recommended. Shrimp and fish are a part of the meals as well.

It looks tasty, and sirtuins are indeed involved in a wide range of cellular processes, including metabolism, aging and circadian rhythm. The diet is also based in part on calorie restriction. The nutritionists behind this diet say that this diet "influences the body's ability to burn fat and that it speeds up the metabolic system."

Diet Under the Microscope

What do we think about the diet? The conclusion is almost nothing, from a scientific point of view. Sirtuins help control fat and glucose metabolism in response to energy-level changes. These may also play a part in strengthening longevity in the impact of calorie restriction. This can be through the influence of sirtuins on aerobic (or mitochondrial) metabolism, reducing varieties of reactive oxygen (free radicals), and increasing antioxidant enzymes.

Furthermore, research shows that transgenic mice with high SIRT6 levels live much longer than normal wild mice and that changes in the expression of SIRT6 may be related to the aging of certain human skin cells. Also, SIRT2 has been shown to slow metazoan aging (in yeast).

All of this sounds impressive, and the diet has received rave reviews and criticism, but none of those comments are compelling evidence that the SirtFood diet has similar effects on humans. It would be a massive extrapolation to conclude that studies on mice, yeasts and human stem cells carried out in the laboratory can be replicated in real life because there could be parasitic and confusing variables.

The Science of Weight Loss

No doubt the plan will work for some people. But the scientific evidence confirming the effectiveness of any diet is something completely different. Of course, the ideal study to consider that a diet is effective for losing weight (or for any other result, such as aging) will require a sufficiently large sample, representative of the population, and a random distribution of the treatment/diet to be followed. And a control group. The results will be recorded over an adequate period of time with strict control of the confounding variables, like any other behavior that can positively or negatively affect the results (for example, smoking, exercise, etc.).

This research would be limited by methods like self-assessment and memory, but which would help discover the effectiveness of this diet. Research of this nature does not exist, however, and one must, therefore, be very careful in the interpretation of basic science, then, human cells in a culture dish most certainly react differently from cells in a living person.

And the doubt about this diet is even greater when you consider some of the statements made about it. Weight losses of 3.20 kg in one week are unrealistic and unlikely to reflect changes in fat in the body. During the first three days, individuals who follow this diet consume around 1,000 kcal per day, around 40% to 50% of what a person needs. This will result in a rapid loss of glycogen (the stored form of carbohydrates) from muscles and the liver.

But for every gram of glycogen stored, we also store about 2.7 grams of water, and the water is heavy. Thus, for all the glycogen lost, we also lose the water that accompanies it and, therefore, the weight. In addition, diets that are too restrictive are very difficult to follow and often result in an increase in appetite-stimulating hormones, such as ghrelin. The weight (glycogen and water) will return to normal if hunger takes over and prevails.

In general, applying the scientific method to study nutrition is difficult. It is often not possible to carry out truly valid placebo-controlled studies, and the health outcomes of interest to researchers spanned several years, making research very difficult. Studies of large populations depend on sometimes simple and naïve data collection methods such as memory and self-assessment, which often produce unreliable data. Against this parasitic noise, nutrition research must do a difficult job.

Is there a Quick Fix?

Regrettably no! The headlines that cause a stir in magazines and often hyperbolic portrayals of scientific data about the endless controversy over what we should eat and how much more fuel our obsession with the quick fix or magic treatment, which is an endemic social problem in itself.

For the reasons mentioned above, the SirtFood Diet should be categorized, at least from a scientific point of view, into the bogus diet group. Or say the contrary from the facts we have is at best fallacious and at worst misleading and damaging or people who want to improve their wellbeing and lose weight. This diet is unlikely to offer benefits to a world facing a diabetes crisis, concealed under the shadow of obesity. Like many have already clearly noted, special diets do not work, and diet is not, in general, a public health option for communities where so many people are overweight.

The best approach now is to change your behavior, mixing it with political and environmental power that helps to improve physical activity and some form of control over what you consume. It's not a fast fix, but it does work!

CHAPTER THREE

Sirtuins and Weight Loss

What is the first thing that comes to mind when you hear the word "diet?" Most people think of losing weight and food restriction. There are hundreds of methods to lose weight in the market, and if you have a constant battle with your weight, you have probably heard of the majority, we go on a new diet every week - or almost.

We know that diets, such as 5:2 or "fasting" activate the so-called "skinny gene" ... it works for some people, but in the end, if you are hungry, you will probably be in a bad mood and even suffer attention deficit that is not only a problem at work, but it can be dangerous. Personally, I have nothing against 5:2 as that can be very useful for people who have a hard time with a daily caloric restriction for a sustained period of time. Very restrictive diets have a very high abandonment rate, and the dieter gets fed up easily and leaves.

The name already reveals what the diet is basically focused on: all those foods which stimulate the fat-burning enzyme sirtuin and reduce the extra pounds are called sirt foods.

The scientific knowledge that certain plant compounds induce the function of the body's own sirtuins, as well as fasting, is the basis of the Sirt Food Diet. For instance, Aidan Goggins and Glen Metten, the nutritionists and writers of the same-named diet bestseller, are persuaded that a diet based on the Sirtuin concept leads not only to a dream figure but also to better well-being so long as you blend the appropriate foods and outsource your metabolism in that way.

Examples of the secondary plant substances known to be sirtuin activators are allicin, which gives garlic its traditional aroma, capsaicin found mainly in chilies, or curcumin, to which the turmeric owes its yellow color.

Sirtfoods not only effectively improve fat burning, but they also protect the body from cell damage, heart or cancer diseases and delay the overall aging cycle. Sirtuins also help users avoid traditional cravings, encourage muscle building and cellular health and reinforce the immune system. What makes a sirtuin diet so much easier: Sirtfoods are not rare or bland at all. Those are daily foods, particularly fruits and vegetables. Red wine and chocolate are allowed, by the way, too.

The Most Important Sirt Foods for Losing Weight

First, the basis: All foods in the Sirtfood diet are plant-based. The precious sirtuins have not only fruit and vegetables, but also spices and herbs. Apples, blueberries, raspberries and citrus fruits, as well as broccoli, cabbage, tomatoes, arugula, onions, celery, garlic, parsley, chili, turmeric, walnuts, and cashew nuts, are common Sirt foods. Sirtuins are also found in 85 percent cocoa dark chocolate and red wine. Such foods are balanced by validated low carb diet sources of protein such as soy products, white meat or eggs.

Physical Activity and the Sirt Food Diet

During the first seven days of the Sirt Food Diet, it is recommended to avoid intensive, exhausting physical activity due to the low caloric content of the diet - a moderate intensity effort is recommended. The authors of the Sirt Food Diet emphasize that when combined with regular exercise, we can achieve better results. They recommend engaging in physical activity five times a week for 30 minutes, maintaining moderate intensity.

During the first part of the diet when your calorie intake is reduced, it would be sensible to stop or reduce exercise while your body adapts to fewer calories. If you feel tired or have less energy than usual, don't work out. Instead, make sure you stay focused on the principles that apply to a healthy lifestyle, such as including adequate daily fiber, protein, and fruit and vegetables.

With an estimated 650 million obese adults worldwide, discovering healthy eating and exercise routines that are doable is crucial, not depriving you of all you love and not forcing you to work out all week. That's the Sirtfood Diet. The idea is that certain foods activate the 'skinny gene' pathways, usually activated by fasting and exercise. The good news is that certain food and drinks, including dark chocolate and red wine, contain chemicals called polyphenols that activate genes that mimic exercise and fasting effects.

When exercising, it's important to consume protein one hour after your workout. After the workout, protein strengthens muscles, decreases soreness and improves recovery. There are a variety of recipes that include protein perfect for post-exercise consumption, such as sirt chili con carne or turmeric chicken and kale salad. If you want something lighter, try the sirt blueberry smoothie and add some protein powder to the benefit. The type of fitness you do will be up to you, but home workouts will allow you to choose the types of exercises that suit you and are short and convenient.

The Sirtfood Diet is a great way to change eating habits, lose weight, and feel healthier. Although the initial few weeks will challenge you, it is important to test which foods are better for eating and which delicious recipes match you. In the first few weeks, be kind to yourself while your body adapts and exercise gently if you choose to do it. If you are already someone doing moderate or intense exercise, you may be able to carry on as normal or manage your fitness in accordance with the change in your diet. As with every change in diet and exercise, it's about the person and how hard you can push yourself.

Foods that Activate the Lean Gene

Chili

Luteolin and myricetin are the two main nutrients that, in the case of chili peppers, activate sirtuin (lean gene).

Cocoa

Who does not like chocolate? Add pure cocoa to your diet. It contains epicatechin, a phytonutrient from the polyphenol family that also has a strong antioxidant action.

Coffee

Coffee is known to be a good ally in weight loss, as it speeds up metabolism. But how does it activate the lean gene? Caffeic acid and chlorogenic acid are the two compounds that activate sirtuin in coffee.

Green Tea

Green tea is a drink with numerous health benefits. Its main feature is to speed up the metabolism, causing the body to spend more energy. This drink is considered Sirt and contains epigallocatechin gallate, a potent activator of sirtuin.

Celery

Celery stimulates appetite, prevents fluid retention, protects against osteoporosis, and fights heart disease.

To activate the lean gene, just use the stem and leaves entirely. They are rich in apigenin and luteolin, two substances that awaken sirtuin.

Arugula

It helps protect against cancer and is rich in antioxidants essential to health. Arugula has several nutrients that activate sirtuin, namely quercetin, and kaempferol.

Saffron

Saffron increases the body's antioxidant capacity, reduces the risk of heart disease, prevents cancer, and helps fight degenerative diseases like Alzheimer's. Many of these benefits arise due to the presence of curcumin, the nutrient that activates sirtuin.

Red Wine

Resveratrol and piceatannol are the two main activators of sirtuin in red wine. Drinking a glass of red wine every day is very good for your health, provided it's done with moderation and balance. In addition to being an ally against weight loss, it helps and improves cognitive processes, treats gum infections, combats tiredness, and reduces cholesterol.

Strawberries

Strawberries contain fisetin, which stimulates energy expenditure and weight loss.

Purple Onion

Purple onion is rich in quercetin, one of the activating compounds of the "lean gene."

Extra Virgin Olive Oil

Olive oil was already mentioned by Hippocrates as "the cure for all ills," well over 2,000 years before modern science proved its wonderful benefits.

When it comes to olive oil, it is essential to buy extra virgin olive oil.

Virgin oil is obtained only through the mechanical pressing of the fruits and in conditions that do not lead to the deterioration of the oil. In this way, you can be sure of the quality of the product and its polyphenol content.

"Extra virgin" refers to the first pressing of the fruits ("virgin" refers to the second), which provides a product of better quality and taste.

Rocket Salad

The rocket was cultivated for the first time in Ancient Rome, where he was considered an aphrodisiac.

Two types are widespread: the salad rocket and the wild rocket. Both are two excellent sirtuin activators.

Turmeric

Turmeric, also known as "Indian solid gold," is believed to be one of the factors contributing to the lower percentage of cancer patients in India compared to western countries.

Studies have shown that a special type of curcumin helps improve cholesterol levels and control blood sugar levels, as well as reducing inflammation.

Turmeric has also proven to be an excellent natural pain reliever in cases of osteoarthritis of the knee.

Foods that Increase Physical Performance

Coffee

Caffeine is one of the most effective nutritional strategies in physical performance for increasing muscle strength and providing energy. The recommendations point to 5-6 mg. per kg. of body weight. However, there are some undesirable side effects: gastrointestinal changes, agitation, and difficulty concentrating. Lower doses of 3 mg./kg. weight seem to contribute to the same results in terms of performance on the physical task, but without suffering from side effects.

Example: For an individual weighing 70 kg., approximately 200 mg. of caffeine (2 espressos or a long coffee) is advised. The desired effects usually occur 40-80 minutes after drinking, so coffee intake should be monitored accordingly.

Sugar

The benefits of increased physical performance and a decreased feeling of effort have been demonstrated when sugars (glucose and fructose or dextrin maltose) are placed in the mouth for approximately 5 seconds. Studies point to an increase of approximately 3% in the physical response and in the reduction of the total time with doses of 6% of glucose or maltodextrins diluted in 25 ml for each 12.5% of exercise (in 1h every 7.5 minutes) and with a fast of at least 3 hours. But, be careful, this approach is only recommended for medium duration exercises (approximately 1 hour). Whenever the exercise exceeds this time, the intake of sugars (drink or energy bar or gel) must be monitored.

Water

Adequate hydration improves physical performance and provides cerebral stability. Water intake must be done in the hours before the workout. Do not consume salty foods at the last meal before exercise, as they increase the feeling of thirst.

7-Day Meal Plan

Monday:
- Breakfast: Cake pineapple and goji berries for a day, coffee, 1 yogurt 0%
- Lunch: Chicken skewers with spices and fresh pasta, egg cream without complex

- Snack: Small Swiss cake, 1 black tea with lemon
- Dinner: Minced beef from Provence, 1 bowl of 0% Fromage blanc

Tuesday:

- Breakfast: 1 Dukan pancake, 1 egg, 1 slice of defatted ham, 1 coffee
- Lunch: Thousand salmon and zucchini leaves, 1 flavored 0% yogurt
- Snack: Rhubarb egg cream, 1 peach green tea
- Dinner: Chicken breast with garden vegetables, 1 bowl of 0% cottage cheese

Wednesday:

- Breakfast: 1 coffee, Muesli, 0% cottage cheese
- Lunch: Spinach salmon fritters, 2 small Swiss 0% with sweetener
- Snack: Blueberry Sunshine, 1 coffee
- Dinner: Scallops with 'honey', konjac tagliatelle, 1 iced Cappuccino

Thursday:

- Breakfast: 1 bowl of tea, soft chestnut
- Lunch: Zucchini-salmon amuse-bouche, chicken fricassee with mushrooms and asparagus

- Snack: French toast and applesauce, 1 coca 0%
- Dinner: Turkey couscous and spicy beef meatballs, 1.5 special coconut cookie

Friday:

- Breakfast: Coffee, 2 slices of whole meal bread, turkey breast, 1 fresh square 0%
- Lunch: Roasted pepper soup, Indian chicken supreme, 1 0% yogurt
- Snack: Sublime rhubarb tarte tatin, 1 red fruit tea
- Dinner: Cannelloni ricotta-spinach, melon mousse

Saturday:

- Breakfast: Tea, 2 flavored yogurts, 2 slices of whole meal bread with syrup jam
- Lunch: Carbonara pasta in my own way, 40g of parmesan, 1 Bio-Flan with coffee
- Snack: Cheesecake with silky tofu and konjac pearls with blueberries
- Dinner: Bib and butternut squash in Parmentier style bistro, 1 bowl of 0% cottage cheese, 1 kiwi cut into it

Sunday:

- Breakfast: Mint tea, 1 sweet cinnamon pancake

- Lunch: Gala meal
- Snack: 1 apple, 2 slices of whole meal bread, 40g of reblochon cheese
- Dinner: Bar en papillote and vegetables, carrot cake

CHAPTER FOUR
Sirt Food Diet: Myth or Truth

Chocolate and red wine as magic weapons in the fight against extra pounds - that sounds too good to be true. But the new Sirt Food Diet isn't that simple.

The basic principle of this nutritional trend is based on the long-propagated low carb diet. It is important to reduce carbohydrate intake in favor of valuable proteins.

Who Came Up with this?

The two nutritionists Aidan Goggins and Glen Matten, wanted to develop a diet that balanced enjoyment and health. The success of their bestseller, "The SIRT Food Diet: A revolution in health and weight loss," was published in German under the title "The Sirtuin Diet - Young and slim with pleasure: How to lose more than 3 kilos in 7 days" proves you right. In the long-term, a diet that focuses on overcoming and permanent renunciation has no chance. The food we eat has to taste good first and foremost. Only then can we stick to a diet and get used to the new concept permanently. And only then can the fight against obesity be won.

How Does the Sirtuin Diet Work?

Sirtuin sounds spacy. The term hides a special protein, more precisely an enzyme that stimulates fat burning in our body. Certain foods have a particularly high sirthrin content. The goal is to turn many of these sirtuin-rich slimming agents into delicious meals. There is a really promising side effect of this diet program for free: Sirtuin increases the number of antioxidant enzymes in the body. Antioxidants are known to have a rejuvenating effect. They can also protect us effectively against free radicals and thus against many diseases.

What Is the Catch?

Admittedly, there is a drop of bitterness. The calorie intake is quite limited. In the first week, it should be limited to 1,000 kcal per day, from the second week to 1,500 kcal. With so little calorific value, weight can be lost without the miracle cure Sirtuin.

Nevertheless, it can be worth giving the Sirtuin Diet a chance. The recommended foods are undoubtedly very healthy. A change in diet that takes us away from fast food and denatured foods is always a good idea. Losing a few extra pounds is also a health benefit. It's better not to overdo it with the red wine.

Are Sirt-Foods New Superfoods?

It cannot be denied that sirt-foods are healthy products. They are often rich in nutrients and full of healthy plant-based compounds. In addition, research has linked many of the foods recommended in the Sirt Food Diet with health benefits such as preventing the occurrence of lifestyle diseases. In fact, most sirt-foods have health benefits for people.

However, evidence of health benefits from increasing Sir protein levels is preliminary. Although studies in mice and human cells in vitro have shown positive results, no studies have been conducted in humans where the effect of increasing sirtuin levels has been studied. Therefore, it is unknown whether increasing the level of Sir proteins in the body will lead to a longer lifespan and a lower risk of certain diseases.

Is the Sirtuin Diet Healthy?

Sirtfoods products are almost all healthy choices. They may even bring some health benefits due to their antioxidant or anti-inflammatory properties. However, eating only a few particularly healthy foods will not meet all of your body's nutritional needs. A sirtuin diet is unnecessarily restrictive and doesn't offer clear, unique health benefits over any other type of diet.

What's more, eating only 1,000 calories a day is not usually recommended without medical supervision. Even consuming 1,500 calories a day may be too restrictive for many people. Due to low calories and restrictive dietary choices, the diet can be difficult to follow for the whole three weeks. Add to this the high initial costs of buying a juicer, books, and some rare and expensive ingredients, as well as the time costs of preparing specific meals and juices, and this diet becomes unfeasible and unbalanced for many people.

Does the Sirtuin Diet Work?

The authors of the Sirtuin Diet boldly claim that this diet can increase weight loss, incorporate a "lean gene" into the body, and prevent disease. The problem is that there is not much evidence to support these claims. So far, there is no convincing research that the Sirt Food Diet has a more beneficial effect on the loss of unnecessary kilograms than any other diet with limited calorie content. Although many of the foods promoted have health properties, no long-term human studies have been conducted to determine if a diet rich in sirtfoods has specific health benefits.

Do You Need to Take Dietary Supplements on a Sirt Food Diet?

A properly balanced Sirt Food Diet, especially in the second phase, will cover our need for essential nutrients. The authors of the Sirt Food Diet recommend that people living in Central Europe take the supplement vitamin D in the autumn and winter. On the other hand, vegans taking sirtfood diets should still remember about vitamin B12 supplementation and may also consider omega-3 supplementation.

When It Is Appropriate to Do it

The Sirt Food Diet can be done at any time, and in almost any type of physical condition, because not excluding any type of food, there is no danger of nutritional deficiency.

When there Is No Need to Do It

The recommended foods may interact with some drugs, so in case of drug therapy, it is good first to hear the opinion of a nutritionist.

What to Drink on a Sirt Food Diet?

On a Sirt Food Diet, it is recommended to drink enough fluids every day, which for the Polish population are 2 liters a day for women and 2.5 liters for men. The authors of the Sirt Food Diet encourage us to reach for:

- Water
- Green tea
- March
- White tea
- Black tea
- Red wine (1 glass with a meal - 2-3 times a week

CHAPTER FIVE

Sirt Food Breakfast Recipes

Honey Cake with Orange Cream

Ingredients

- 200 g apples (1 apple)
- 3 eggs
- 200 g honey
- 100 ml rapeseed oil
- 1 pinch salt
- 1 tsp. cinnamon (preferably Ceylon cinnamon)
- 300 g wholemeal flour
- ½ packet baking powder
- 100 ml apple juice
- 70 g margarine
- 200 g organic orange (1 organic orange)
- 50 g unsulphured raisins
- 100 g carrots (1 carrot)
- 25 g unshelled almond kernels

Directions: 25 min

1. Wash, quarter, core, and grate the apple finely on a grater.
2. Separate eggs. Put the egg yolks with honey in a bowl and stir in a hand mixer until creamy.

3. Gradually fold in the rapeseed oil. Finally, add a pinch of salt, grated apple, and cinnamon.
4. Sieve wholemeal flour with baking powder to the egg yolk cream and stir in alternately with the apple juice.
5. Put the egg whites in another bowl. Whisk until stiff with a hand mixer and fold under the dough.
6. Grease a tin (29x10.5 cm) with 1 tbsp. margarine.
7. Pour in the dough, smooth it with a rubber spatula and bake in the preheated oven at 180 ° C (fan oven: 160 ° C, gas: level 2-3) on the middle shelf for 50-60 minutes.
8. In the meantime, wash the orange in hot water, rub it dry and rub the peel finely.
9. Halve the orange and squeeze out the juice. Mix the juice in a small bowl with the raisins and let it soak a little.
10. Wash, clean, peel and cut the carrot into large pieces. Finely chop with the almonds in the Blitzhacker.
11. Remove the raisins from the juice. Mix in the remaining carrot and almond mixture with the remaining margarine.
12. Stir in the grated orange peel and 2-3 tablespoons of orange juice. Pour into a small screw-top jar with a lid and chill.
13. Take the honey cake out of the oven and let it cool completely on a wire rack. Cut off 1 slice per serving and spread with 1 tablespoon of orange cream.

Nutritional values | Calories 248

Protein 4 g, Fat 13 g, Carbs 27 g, Added Sugar 10 g, Fiber 3 g

Banana Bread with Walnuts

Ingredients

- 300 g wheat flour type 1050
- 1 packet baking powder
- ½ tsp. salt
- Nutmeg
- 150 g walnut kernels
- 500 g ripe bananas (3 ripe bananas)
- 1 vanilla bean
- 1 apple
- 80 g butter
- 50 g coconut sugar
- 1 egg

Directions: 30 min

1. Sift the flour with baking powder and salt into a mixing bowl. Add a little nutmeg directly.
2. Finely chop walnuts in a blitz chopper or with a large knife and add to the flour mixture.
3. Peel the bananas, cut them into small pieces, put them in a bowl, and puree them with a hand blender or finely mash them with a fork.

4. Cut the vanilla pod lengthways, scrape out the pulp, and stir in the banana puree.
5. Wash the apple, grate, and stir in a bowl with butter and coconut blossom sugar in a bowl with the whisk of a hand mixer, stir in the egg. Then, alternately, gradually pull the banana puree and flour mixture under the butter mixture.
6. Grease a small baking tin (approx. 8x22 cm) if necessary. Pour in the dough and smooth it out with a rubber spatula.
7. Bake in the preheated oven at 175 ° C (fan oven: 150 ° C, gas: speed 2) on the middle shelf for 50–60 minutes. Put a wooden skewer in the middle of the cake: if it stays clean when pulled out, the cake is done; otherwise, continue baking for a few more minutes.
8. Take the finished bread out of the oven, let it cool in the baking pan for 5 minutes, then turn it over.

Nutritional values | Calories 176

Protein 4 g, Fat 9 g, Carbs 19 g, Added Sugar 2.4 g, Fiber 1.8 g

Currant and Banana Croissants with Ground Almonds

Ingredients

- 135 ml milk (1.5% fat)
- ¼ cube yeast
- 125 g wheat flour type 1050
- 125 g wheat flour type 550

- 20 g liquid honey (1 heaped tablespoon)
- 30 g yogurt butter (2 tablespoons; room warm)
- 1 pinch salt
- 250 g red currants
- 50 g dried bananas
- 20 g ground almond kernels (2 tbsp.)

Directions: 40 min

1. Put the milk in a small saucepan, remove 1 tbsp., and set aside in a small bowl. Warm the remaining milk slightly (35 to 40 degrees Celsius). Remove from the heat, crumble the yeast, stir thoroughly and cover, and let rise for about 10 minutes.
2. In the meantime, put both flours in a mixing bowl, mix, press a well in the middle with a tablespoon and add honey. Divide the butter into small portions with a teaspoon and put on the floured rim. Add a pinch of salt.
3. Pour the yeast and milk mixture into the recess and knead everything with the kneading hooks of the hand mixer to a smooth, shiny dough. Cover with a kitchen towel and leave in a warm place for about 40 minutes.
4. In the meantime, wash the currants for the stuffing, drain and rub off the panicles with a fork.
5. Chop dried bananas very finely, mix in a bowl with the ground almonds and currants.

6. Place baking paper on a baking sheet. Put the yeast dough on the lightly floured work surface and knead with your hands for about 2 minutes.
7. Roll out the dough into a circle of approx. 32 cm. Cut into 12 equal triangles with a pizza wheel like a cake.
8. Put some water in a small bowl and thinly brush the edges of the triangles.
9. Put some of the fillings on the bottom wide third of each triangle. Roll up the triangles from the occupied side to the tip and form a croissant.
10. Place the croissant on the baking sheet with the rolled-up tip down and cover and leave to rise in a warm place for about 20 minutes.
11. Brush the croissants with the remaining milk and bake at 200 ° C (fan oven: 180 ° C, gas: level 3) in the preheated oven for about 25 minutes.
12. Take out the baking tray, pull the currant and banana croissants with the baking paper onto the upside-down oven rack and let cool.

Nutritional values | Calories 126

Protein 3 g, Fat 3 g, Carbs 20 g, Added Sugar 2 g, Roughage 3 g

Pea Protein Sandwiches

Ingredients

- 370 g frozen peas
- 225 g oatmeal
- 75 g peeled hemp seeds
- 1½ tsp. fennel seeds
- 1 tsp. coriander seeds
- 1 tsp. caraway seeds
- 2 tsp. baking powder
- 2 tsp. seasoned salt
- 3 eggs (l)
- 2 carrots
- 250 g quark (20% fat in dry matter) salt pepper
- ½ bundle chives (10 g)

Directions: 20 min

1. Let the peas thaw. Grind oatmeal and hemp seeds to flour in a blender. Finely grate the fennel, coriander seeds and caraway seeds in a mortar and add to the flour. Add baking powder and salt and mix.
2. Put the peas and eggs in the blender and chop until you get a smooth dough. Line a baking tin with baking paper and add the dough.
3. Peel the carrots, quarter them lengthways and place them on the dough. Bake in a preheated oven at 180 ° C (fan oven 160 ° C; gas: levels 2–3) for 55–60 minutes. Then let it cool down for about 30 minutes.

4. In the meantime, mix the curd with salt, pepper and 2 tablespoons of water. Wash chives, shake dry and cut into rings. Cut bread into slices, brush with the curd cheese and garnish with chives.

Nutritional values | Calories 212 kcal

Protein 9 g, Fat 13 g, Carbs 8 g, Added Sugar 2 g

Sweet Pumpkin Buns

Ingredients

- 300 g Hokkaido pumpkin
- 50 ml of orange juice
- 450 g spelled flour type 1050
- 1 cube yeast
- 70 g whole cane sugar
- 150 ml lukewarm milk (3.5% fat)
- 1 vanilla bean
- 1 egg (m)
- 80 g room temperature butter
- ½ tsp. cinnamon
- 1 msp. cardamom powder
- 1 pinch salt
- 1 egg yolk

Directions: 30 min

1. Wash, core and dice the pumpkin. Place in a saucepan with orange juice and cook gently on a low heat for approx. 15 minutes. Puree and let cool.
2. In the meantime, put the flour in a bowl and press a hollow in the middle. Crumble the yeast and add 1 tsp. whole cane sugar and milk to the well. Cover and let rise for 10 minutes.
3. Slit the vanilla pod lengthways and scrape out the pulp. Add the vanilla pulp, remaining sugar, egg, butter, cinnamon, cardamom, 1 pinch of salt, and cooled pumpkin puree to the batter. Knead everything into a smooth dough and cover and let rise for 1 hour.
4. Divide the dough into eight equal pieces and shape them into rolls. Place on a baking sheet covered with baking paper. Whisk the egg yolk with water and brush the rolls with it.
5. Bake the rolls in a preheated oven at 180 ° C (forced air 160 ° C, gas: level 2–3) in 20–30 minutes until golden brown.

Nutritional values Calories 349

Protein 10 g, Fat 13 g, Carbs 48 g, Added Sugar 9 g, Fiber 5.8 g

Avocado Smoothie with Yogurt and Wasabi

Ingredients
- 1 bunch coriander
- 1 spring onion
- 2 avocados

- 1 lime
- 1 tsp. wasabi paste
- 500 ml kefir
- 450g yogurt (0.3% fat)
- 2 handfuls ice cubes
- Salt
- Pepper

Directions: 15 min

1. Rinse the coriander, shake it dry, and pluck the leaves. Clean, rinse, drain the spring onions, and cut them into rings.
2. Halve and stone avocados. Remove the pulp from the bowls with a tablespoon and place it in a blender or a tall container with coriander and spring onion rings.
3. Squeeze the lime. Add three tablespoons of juice, wasabi paste, kefir, and yogurt to the avocado.
4. Puree everything in the blender or with a hand blender, gradually adding the ice cubes. Season the avocado smoothie with salt and pepper and pour it into glasses.

Nutritional values | Calories 172

Protein 7 g, Fat 11 g, Carbs 7 g, Fiber 1.5 g

Goat Cheese Omelet with Arugula and Tomatoes

Ingredients

- 4 protein (s)

- 2 eggs (s)
- 1 small handful arugula
- 2 tomatoes
- 1 tsp. olive oil
- Salt
- Pepper
- 50 g young goat cheese

Directions: 15 min

1. Separate 4 eggs and put the egg whites in a bowl (use egg yolks elsewhere). Add the remaining 2 eggs and whisk everything with a whisk.
2. Wash the rocket, spin dry, and chop it roughly with a large knife.
3. Wash the tomatoes, cut out the stem ends in a wedge shape, and cut the tomatoes into slices.
4. Heat a coated pan (24 cm) and spread with the oil.
5. Add the whisked egg mixture. Season with salt and pepper.
6. Bake slightly over medium heat (the egg should still be a little runny) and turn using a plate.
7. Crumble goat cheese over the omelet with your fingers. Put the omelet on a plate, top with tomato slices and sprinkle the rocket. Whole grain toast goes well with this.

Nutritional values | Calories 430

Protein 43 g, Fat 23 g, Carbs 10 g, Fiber 2.5 g

Baked Quark Toasts with Orange Fillets

Ingredients

- 1 vanilla bean
- 250 g buttermilk curd
- 3 tbsp. maple syrup
- 1 large organic orange (or 2 small ones)
- 2 eggs
- 225 ml milk (1.5% fat)
- 1 pinch salt
- 8 slices whole-grain toast
- 2 tbsp. rapeseed oil

Directions: 35 min

1. Halve the vanilla pod lengthways and scrape out the pulp with a knife.
2. Mix the curd with the maple syrup and the vanilla pulp until smooth.
3. Rinse orange hot and rub dry. Rub the peel finely and stir in the curd.
4. Peel the orange so that everything white is removed. Cut out the fruit fillets between the cuticles, place in a small bowl and set aside.
5. Whisk eggs, milk, and salt in a shallow baking dish. Turn the slices of bread in portions.

6. Spread a large coated pan with 1/2 tablespoon of oil and heat. Bake 2 slices of bread on each side for 1–2 minutes until golden brown. Continue until all the oil is used up, and all the toasts are baked. Serve with quark and orange fillets.

Nutritional values | Calories 338

Protein 18 g, Fat 11 g, Carbs 40 g, Added Sugar 7 g, Fiber 2.5 g

Sweet Millet Casserole with Clementine

Ingredients

1. 150 g golden millet
2. 1 vanilla bean
3. ½ organic lemon
4. ½ tsp. cinnamon powder
5. 2 tbsp. raw cane sugar
6. 380 ml oat drink (oat milk)
7. 2nd tangerines
8. 3rd eggs
9. 1 pinch salt
10. 300 g lactose-free curd cheese (20% fat)
11. 1 tbsp. butter
12. 2 tbsp. sliced almonds

Directions: 30 min

1. Wash millet in hot water. Halve the vanilla pod lengthways and scrape out the pulp with a knife. Rinse half of the lemon in hot water, pat dry and rub the zest finely. Squeeze out the juice.
2. Put millet, vanilla pulp, cinnamon, sugar and oat drink in a saucepan. Bring to the boil and simmer over medium-high heat for about 7-10 minutes, stirring occasionally. Remove from the heat and let soak for 10 minutes without a lid.
3. Meanwhile, peel the tangerines and cut them into thick slices. Separate eggs. Beat egg whites with 1 pinch of salt until egg whites are stiff. Mix the egg yolks with lemon zest, lemon juice and quark and add to the millet mass. Carefully fold in the egg whites.
4. Butter the baking dish. Pour in the millet curd mixture, smooth out and top with the mandarins. Bake in a preheated oven at 180 ° C (fan oven 160 ° C; gas: levels 2–3) for 40–50 minutes. Sprinkle millet casserole with grated almonds to serve.

Nutritional values | Calories 420

Protein 20 g, Fat 15 g, Carbs 49 g, Added Sugar 7.5 g, Fiber 3.6 g

Cloud Bread

Ingredients

1. 3 eggs
2. 200 g cream cheese (60% fat in dry matter)
3. 1 tsp. baking powder

4. Salt 50 ml
5. Milk (3.5% fat) ½ tsp.
6. Medium-hot mustard pepper
7. ½ bundle chives (10 g)
8. 100 g lamb's lettuce
9. ½ bundle radish
10. 1 handful red radish cress

Directions: 10 min

1. Separate the eggs and mix the egg yolks with 100 g cream cheese and baking powder. Beat the egg whites with a pinch of salt until stiff and fold in portions under the cream cheese cream.
2. Spread the dough in 8 flat portions on a baking sheet covered with baking paper and bake in a preheated oven at 150 ° C (convection 130 ° C; gas: setting 1-2) for about 20 minutes. Take out and let cool for about 10 minutes.
3. Mix the remaining cream cheese with milk, mustard, a little salt, and pepper. Wash the chives, shake dry, cut into rolls and stir into the cream.
4. Clean lamb's lettuce, wash and shake dry. Clean, wash and cut radishes into thin slices. Wash the cress and drain well. Spread some cream on the underside of 4 cloud pieces of bread each, arrange lamb's lettuce, radish slices, remaining cream and cress on top and place one cloud bread each on top as a lid.

Nutritional values | Calories 71

Protein 4 g, Fat 6 g, Carbs 1 g

Cottage Cheese with Raspberry Sauce

ingredients

- 400 g flour-boiling potatoes (2-3 potatoes)
- 300 g raspberries
- 2 tbsp. honey
- ½ vanilla bean
- 250 g low-fat quark
- 50 g coconut sugar
- 150 g spelled flour type 1050
- 1 egg
- Cinnamon
- 15 g butter (3 tsp.)
- 30 g planed almond kernels (2 tbsp.)

Directions: 5 min

1. For the quark legs, peel, wash, chop the potatoes and cook gently in boiling water in about 15 minutes over medium heat. Then pour off and let cool for 10 minutes.
2. In the meantime, wash the raspberries carefully and puree them with a hand blender. Push the pulp through a sieve, mix with honey and keep cool.

3. In the meantime, halve the vanilla pod lengthways and scrape out the vanilla pulp with a knife.
4. Press potatoes through a potato press into a bowl. Add the curd, sugar, flour, egg, vanilla pulp and 1 pinch of cinnamon to the potatoes and knead everything into a smooth dough; if it is too moist, add some flour. Form 12 small cookies out of the dough.
5. Fry the quark balls in succession. Heat 1 teaspoon butter in a pan. Add 4 balls and bake golden brown on each side in about 3-4 minutes; Bake was remaining cups in the same way.
6. Toast the almonds in a hot pan without fat over medium heat for 3 minutes. Arrange the quark drumstick with the raspberry sauce and almonds.

Nutritional values | Calories 61

Protein 3 g, Fat 5 g, Carbs 1 g

Oriental Porridge with Oranges and Figs

Ingredients
- 400 ml oat drink (oat milk)
- 75 g pithy oatmeal
- 25 g tender oatmeal
- 2 tbsp. maple syrup
- 2nd cardamom capsules
- 1 tsp. cinnamon
- ¼ tsp. vanilla powder

- 1 pinch salt
- 1 orange
- 1 coward
- 2 tbsp. white almond butter
- 1 tbsp. crushed gold flax seeds (10 g)
- 1 tbsp. chopped pistachio nuts (15 g)

Directions: 15 min

1. Heat the oat drink in a saucepan. Stir in oatmeal, maple syrup, cardamom, cinnamon, vanilla and a pinch of salt and simmer over medium heat for 2-3 minutes. Then allow for swelling for about 5 minutes over low heat.
2. Meanwhile, peel the orange so that all white is removed. Cut out fruit fillets between the cuticles while catching the juice. Wash the fig, pat dry and cut into eighths. Mix the almond butter with orange juice.
3. Fill the oriental porridge in small bowls and serve with orange fillets, fig slices, linseed and pistachios. Drizzle the almond butter over the porridge before serving.

Nutritional values | Calories 533

Protein 15 g, Fat 21 g, Carbs 70 g, Added Sugar 11 g, Fiber 10.3 g

Blackberry and Apple Spread with Lavender Flowers

Ingredients

- 400 g blackberry

- 450 g apples (3 apples; e.g., elstar)
- 250 g gelling sugar 3: 1
- 1 tsp. dried lavender flowers

Directions: 25 min

1. Rinse the appropriate screw-top jar or jars (for a total of 900 ml) with matching lids with boiling water and let them drip upside down on a kitchen towel. Rinse blackberries carefully in a sieve and let them drain.
2. Wash, peel, quarter, and core apples.
3. Halve the length of the apple and cut it into slices.
4. Halve large blackberries if necessary. Weigh blackberries and apples and mix 750 g of them in a large saucepan with the gelling sugar.
5. Bring everything to a boil over high heat and let it boil for at least 3 minutes, stirring constantly.
6. Fold in the lavender flowers, bring to the boil again briefly, and immediately pour into the prepared glasses. Seal, turn around and let stand upside down for 5 minutes. Turn glasses over again and let cool. (The spread stays good closed and stored in a cool place for about six months.)

Nutritional values | Calories 19

Carbs 4 g, Added Sugar 3 g, Fiber 0.5 g

Hearty Smoked Salmon Slices with Cream Cheese and Onion

Ingredients

- ½ lemon
- 75 g cream cheese (13% fat)
- Salt
- Pepper
- 5 stems chervil
- 4 slices whole rye bread
- 1 small red onion
- 100 g smoked salmon

Directions: 10 min

1. Halve and squeeze the lemon. Mix cream cheese in a bowl until creamy. Season with salt, pepper and lemon juice.
2. Wash chervil, shake dry, pluck leaves, chop and stir under the cream cheese.
3. Lightly toast bread slices in the toaster or under the preheated grill. Peel the onion and cut it into fine rings.
4. Brush the bread with cream cheese and top with salmon slices. Spread the onion rings on top and serve the bread.

Nutritional values | Calorie 329

Protein 20 g, Fat 9 g, Carbs 39 g, Roughage 9 g

CHAPTER SIX

Sirt Food Lunch Recipes

Vegan Raw Meatballs

Ingredients

- 250 g fine bulgur (köftelik bulgur)
- 1 clove of garlic
- 2 beef tomatoes
- 1 tsp. salt
- ½ tsp. cayenne pepper
- 1 tsp. cumin
- 1 tsp. paprika powder
- 1 pinch chili flakes
- 40 g paprika market (3 tbsp.)
- 25 g tomato paste (2 level tablespoons)
- 2 spring onions
- ½ bundle parsley (10 g)
- ½ bundle mint (10 g)
- 2 lettuce hearts
- 1 organic lemon
- 5 radishes

Directions: 20 min

1. Pour 250 ml of hot water over the bulgur, stir and allow for swelling for 10 minutes.
2. Peel and finely chop the garlic. Scald the tomatoes with boiling water, quench and skin. Remove the seeds and cut the pulp into fine cubes.
3. Add garlic, tomatoes, salt, spices as well as paprika and tomato paste to the bulgur and knead vigorously for about 5 minutes (preferably with gloves). Let it rest for 20 minutes.
4. In the meantime, clean, wash and cut the spring onions into very fine rings. Wash parsley and mint, shake dry and chop finely. Separate lettuce leaves from the lettuce head, wash and shake dry. Rinse the lemon in hot water, grate dry, cut in half, and slice. Clean, wash and cut radishes into small cubes.
5. Knead parsley, half of the mint and spring onions under the dough. Cut off about 30 walnut-sized pieces and crush them in your hand so that elongated rolls (köfte) are created.
6. Place the kofte in a leaf of lettuce, sprinkle with the remaining mint and radishes and serve with lemon.

Keto Bowl with Mushrooms and Chinese Cabbage

Ingredients

- 300 g Chinese cabbage
- 100 g tomatoes
- 200 g smoked tofu

- 10 g chili pepper
- 400 g mushrooms
- 1 tbsp. rapeseed oil
- Salt
- Pepper
- 30 g macadamia nuts
- 20 g red currants
- 100 g soy-based skyr
- ½ tsp. medium-hot mustard
- 1 tsp. apple cider vinegar
- 1 tbsp. tahini
- 1 stem parsley

Directions: 25 min

1. Clean cabbage, remove outer leaves and stalks, cut into fine strips, wash and drain well. Wash the tomato and cut it into narrow strips together with tofu. Wash, halve, core, and chop the chili. Clean the mushrooms, remove the ends of the stems, and quarter them.
2. Heat the oil in a pan and fry the tofu until golden brown for 4–5 minutes, then remove. Fry the mushrooms over high heat for 2-3 minutes and then remove them. Add the cabbage, tomatoes, and chili to the pan and braise for 1-2 minutes over medium heat. Season with salt and pepper.

3. Meanwhile, chop macadamia nuts and roast them in a small pan without fat, stirring until they smell. Then let it cool.
4. For the sauce, mix the skyr, mustard, vinegar, and tahini with 2 tablespoons of water and season with salt and pepper. Wash, drain and pluck currants. Wash the parsley, shake it dry and pick the leaves.
5. Spread the vegetables and tofu in small bowls, add the sauce and serve sprinkled with nuts, parsley, and berries.

Nutritional values | Calories 450

Protein 35 g, Fat 30 g, Carbs 8 g, Fiber 10.5 g

Spaghetti with Pumpkin and Spinach and Goat's Cream Cheese Sauce

Ingredients

- 400 g whole grain spaghetti
- Salt
- 1 red onion
- 300 g Hokkaido pumpkin pulp
- 300 g spinach leaves
- 2 tbsp. olive oil
- Pepper
- 150 g goat cream cheese (45% fat in dry matter)
- 45 g pumpkin seeds (3 tbsp.)

Directions: 30 min

1. Cook the pasta bite-proof in plenty of boiling salted water according to the package instructions. Then pour off in a sieve and drain.
2. In the meantime, peel, halve and cut the onion into strips. Cut the pumpkin pulp into strips. Clean spinach leaves, wash thoroughly and cut into strips.
3. Heat 1 tablespoon of oil in a pan. Add the spinach and braise for 5 minutes over medium heat until it collapses. Heat the remaining oil in another pan. Braise the onions and pumpkin for 5 minutes over medium heat; then season with salt and pepper.
4. Add cream cheese and about 4–5 tablespoons of water to the spinach and simmer with stirring for 2 minutes, then salt and pepper.
5. Place the pumpkin noodles on a plate, drizzle the spinach sauce over them and serve sprinkled with pumpkin seeds.

Salad Boat with Chickpeas and Tzatziki

Ingredients

- 200 g chickpeas (canned; drained)
- 1 tbsp. sesame oil
- Salt
- Cayenne pepper
- ¼ tsp. turmeric powder
- 1 clove of garlic

- ¼ cucumber
- 200 g Greek yogurt
- 1 tsp. lemon juice
- 200 g red cabbage
- 8 sheets romaine lettuce
- 2 stems parsley
- 50 g feta (45% fat in dry matter)

Directions: 20 min

1. Rinse and drain the chickpeas. Heat oil in a pan. Roast the chickpeas in the medium heat for 5-7 minutes. Season with salt, cayenne pepper, and turmeric.
2. In the meantime, peel and chop the garlic for the Zaziki. Clean and wash the cucumber, grate half, and slice the rest. Mix the garlic and sliced cucumber with the yogurt and lemon juice and season with salt.
3. Clean, wash, and cut the red cabbage into fine strips. Wash lettuce leaves and parsley and shake dry. Roughly chop parsley. Crumble the cheese.
4. Put the chickpeas, red cabbage, and cucumber slices in the lettuce leaves. Sprinkle the salad boat with parsley and feta and serve with tzatziki.

Nutritional values | Calories 200

Protein 9 g, Fat 12 g, Carbs 14 g, Fiber 4.4 g

Stuffed Peppers with Quinoa, Ricotta and Herbs

Ingredients

- 150 g quinoa
- 1 clove of garlic
- 1 shallot
- 2 tbsp. olive oil
- 2 eggs
- 250 g ricotta
- Salt
- Pepper
- 1 tbsp. lemon juice
- 1 bunch mixed herbs (20 g; e.g., thyme, rosemary, oregano, sage)
- 1 map. cayenne pepper
- 4 small red peppers (600 g)
- ¼ bund petersilie (5 g)

Directions: 25 min

1. Rinse quinoa in a sieve with water. Place in a saucepan with twice the amount of water, bring to the boil and cook on a low heat for 10-15 minutes. Then pour off, quench, and drain.
2. Peel and finely chop the garlic and shallot. Heat 1 tablespoon of olive oil in a pan. Braise the garlic and shallot in it for 2 minutes over medium heat.

3. Mix the eggs and ricotta, season with salt and pepper, and lemon juice. Wash herbs, shake dry, pluck leaves, chop finely and add with cayenne pepper to the egg-ricotta mixture. Fold in the quinoa, garlic, and shallot.
4. Wash the peppers, pat dry, cut in half and remove the seeds. Spread a baking dish with the remaining oil. Put the pepper halves in and fill evenly with the ricotta mixture.
5. Bake the filled pepper halves in a preheated oven at 200 ° C (fan oven 180 ° C; gas: setting 3) for about 25 minutes.

Nutritional values | Calories 394

Protein 18 g, Fat 20 g, Carbs 35 g, Fiber 8.6 g

Salmon with Herb and Walnut Salsa

Ingredients

- 1200 g salmon fillet with skin
- Salt
- Pepper
- 6 tbsp. olive oil
- ½ bundle parsley (10 g)
- 4 stems dill
- 60 g walnut kernels
- 1 clove of garlic
- 2 tbsp. capers (30 g; glass; drained)
- Peel and juice from 1 organic lemon

Directions: 20 min

1. Rinse the salmon fillet, pat dry and put it skin-side down on a baking sheet covered with baking paper, season with salt and pepper, and drizzle with 2 tablespoons of oil. Bake in a preheated oven at 200 ° C (fan oven 180 ° C; gas: setting 3) for 12–15 minutes.
2. In the meantime, wash the parsley and dill for the salsa, shake dry and roughly chop both. Roast walnuts in a hot pan without fat over medium heat for 3 minutes; then let cool for 5 minutes. Roughly chop the nuts and put them in a bowl. Peel the garlic. Chop garlic and capers, add both with herbs, lemon zest, and juice to the nuts. Mix everything with the remaining olive oil, salt, and pepper.
3. Take the salmon out of the oven and pull it off the baking sheet with the baking paper. Pour salsa over the salmon and sprinkle with pepper.

Jackfruit Fricassee with Pea Rice

Ingredients

- 250 g brown rice
- Salt
- 250 g frozen peas
- 400 g jackfruit (can; drained weight)
- 1 small onion (40 g)

- 200 g mushrooms
- 3 carrots
- 2 tbsp. rapeseed oil
- 25 g wholemeal spelled flour (2 tbsp.)
- 400 ml vegetable broth
- 250 ml of soy cream
- Pepper
- ½ organic lemon (zest and juice)
- Nutmeg
- ½ bundle chervil

Directions: 45 min

1. Put brown rice in 500 ml of boiling salted water, stir once and let cook in a closed saucepan over low heat for 25–30 minutes until the rice has absorbed the water. Add the peas in the last 5 minutes and finish cooking. Then briefly loosen up the rice in the saucepan and let it swell for a few minutes in the closed saucepan on the switched-off the hob.
2. While the rice is cooking, rinse pieces of jackfruit, drain well and pluck into small pieces. Peel the onion and cut into fine strips. Clean the mushrooms and cut them in slices. Peel the carrots, halve lengthways and cut into slices.
3. Heat oil in a pot. Add the onion and braise for 2 minutes over medium heat. Add pieces of jackfruit and sauté for 5 minutes. Then add mushrooms and carrots and braise for 3 minutes. Dust

everything with flour and pour in the vegetable stock while stirring and simmer on low heat for 10 minutes, adding a little water if necessary.

4. Then stir in the soy cream, season with salt, pepper, lemon zest and juice and freshly grated nutmeg. Heat rice over low heat. Wash chervil, shake dry and chop. Serve fricassee sprinkled with rice and chervil.

Nutritional values | Calories 584

Protein 16 g, Fat 18 g, Carbs 87 g, Fiber 13.1 g

Dill Patties with a Dandelion Dip

Ingredients

- 150 g tender oatmeal
- 20 g dill (1 bunch)
- 1 shallot
- 1 clove of garlic
- 2½ tbsp. olive oil
- 300 g small cucumber (1 small cucumber)
- 50 g gouda (1 piece; 45% fat in dry matter)
- 1 egg
- 20 g crushed flax seeds (2 tbsp.)
- 90 g whole grain bread crumbs
- Salt
- Pepper

- ½ lemon
- 10 g dandelion (3 sheets)
- 250 g yogurt (3.5% fat)

Directions: 45 min

1. Put the oatmeal in a bowl, pour twice the boiling water over it, mix well, and let it swell for about 10 minutes.
2. In the meantime, wash the dill, shake dry and chop finely. Peel the shallot and garlic and chop finely. Heat 1 teaspoon of oil in a small pan. Braise shallot and garlic in it for 3 minutes over medium heat. Clean, wash, and roughly grate the cucumber. Rub Gouda.
3. Mix oatmeal with egg, linseed, herbs, cucumber, cheese, shallot, garlic, and breadcrumbs and season with salt and pepper; add breadcrumbs depending on the consistency.
4. Form the mass into 12 patties and bake one after the other. To do this, heat 1 teaspoon of oil in a pan. Add four patties and fry until golden on each side in 3-4 minutes over medium heat. Fry the remaining patties as well. Let the patties cook for 15 minutes in a preheated oven at 100 ° C (fan oven 80 ° C; gas: setting 1).
5. In the meantime, squeeze half the lemon, wash the dandelions, shake dry and chop. Process together with one tablespoon of lemon juice, yogurt, salt, and pepper into a dip. Serve the patties with the dip.

Nutritional values | Calories 407

Protein 16 g, Fat 19 g, Carbs 43 g, Fiber 7.1 g

Konjac Pasta with Berries

Ingredients

- 1 organic lime
- 2 shallots
- 1 clove of garlic
- 1 red chili pepper
- 200 g celery (3 stalks)
- 2 tbsp. rapeseed oil
- 100 ml vegetable broth
- 200 ml almond cuisine or soy cream
- 30 g pine nuts (2 tbsp.)
- 75 g strawberries (5 strawberries)
- 10 g parsley (0.5 bunch)
- Salt
- Pepper
- 400 g konjac noodles (spaghetti)

Directions: 20 min

1. Rinse the lime hot, pat dry and rub a little peel finely. Halve the lime and squeeze out the juice.
2. Peel the shallots and garlic and dice finely. Halve, chop, wash and chop lengthways. Clean the celery, if necessary untangle, wash and cut into 1 cm thick slices. Roughly chop celery green.

3. Heat oil in a pot. Add the shallots, garlic, chili, celery, and celery greens and sauté for 3-4 minutes over medium heat. Deglaze with broth. Pour in the almond cuisine, bring to the boil and let simmer for 5–6 minutes.
4. In the meantime, roast pine nuts in a hot pan without fat over medium heat for 3 minutes. Clean, wash pat dry strawberries and cut into small cubes. Wash parsley, shake dry and chop.
5. Add parsley to the sauce. Season with salt, pepper, lime peel, and juice. Rinse konjac spaghetti thoroughly and cook in boiling salted water for 2 minutes. Drain, add to the herb-lime sauce, and mix well. Put on a plate, garnish with pine nuts and strawberries.

Nutritional values | Calories 409

Protein 8 g, Fat 36 g, Carbs 9 g, Fiber 11.4 g

Carrot Tagliatelle with Avocado Pesto

Ingredients

- 30 g pine nuts (2 tbsp.)
- 1 ripe avocado
- 2 tbsp. lime juice
- Salt
- Pepper
- 1 pinch chili flakes
- 800 g long carrots (8 long carrots)

- 80 g rocket (1 bunch)
- 3 stems oregano
- 2 tbsp. olive oil
- 75 ml vegetable broth
- 250 g cherry tomatoes

Directions: 30 min

1. Roast pine nuts in a hot pan without fat over medium heat for 3 minutes. Halve the avocado, remove the stone, lift the pulp out of the bowl with a spoon, and place it in a tall mug. Add half of the pine nuts and the lime juice and mash everything finely. Season with salt, pepper, and chili flakes.
2. Clean and peel the carrots and cut them into long narrow strips (tagliatelle) with a peeler. Wash oregano, shake dry, and pluck the leaves.
3. Heat oil in a large pan. Fry the carrot tagliatelle for 4 minutes over medium heat. Deglaze with broth and simmer for 5 minutes on low heat, occasionally turning, until the liquid has boiled. In the meantime, clean, wash and quarter cherry tomatoes.
4. Season the carrot tagliatelle with salt and pepper. Add tomatoes and sauté for 2 minutes.
5. Arrange carrot tagliatelle with the avocado pesto and sprinkle with pine nuts and oregano leaves.

Nutritional values | Calories 237

Protein 5 g, Fat 16 g, Carbs 18 g, Fiber 9.8 g

Strawberry and Avocado Salad with Chicken Nuggets

Ingredients

- 350 g chicken breast
- 1 tbsp. soy sauce
- Pepper
- 1 tsp. sweet paprika powder
- 2 tsp. tomato paste
- 4 handfuls mixed salad (arugula, lollo rosso, lettuce)
- 150 g strawberries
- 1 lemon
- 1 avocado
- 1 egg (l)
- 2 tbsp. wholemeal flour
- 70 g cornflakes (without sugar)
- 3 tbsp. rapeseed oil
- 1 tbsp. honey
- Salt
- ½ red onion
- 5 g ginger
- 2 tsp. sunflower seeds
- 2 tsp. pine nuts

Directions: 55 min

1. Rinse the chicken breast, pat dry, and cut into 6 pieces. Mix in a bowl with soy sauce, pepper, paprika powder, and tomato paste and let marinate for about 10 minutes.
2. In the meantime, clean, wash and spin dry lettuce. Clean, wash pat dry strawberries and cut into small pieces. Squeeze lemon. Halve the avocado, remove the stone, remove the pulp from the skin, dice and mix with half of the lemon juice.
3. Place the egg in a soup plate and whisk with a fork. Put the flour on a second plate. Crumble the cornflakes and put them on another plate. Bread chicken pieces first in flour, then in egg, and then in the flakes. Heat the oil in a pan and fry the chicken on each side over medium heat.
4. Mix a dressing out of honey, remaining lemon juice, salt, and pepper. Peel the onion and ginger, dice the onion and grate the ginger, add both to the dressing. Mix sunflower and pine nuts and roast in a hot pan without fat on a low heat for about 4 minutes.
5. Spread the salad on two plates, pour the strawberries and avocado over it and drizzle with the dressing. Serve with chicken nuggets and sprinkle the salad with the seeds.

Nutritional values | Calories 788

Protein 56 g, Fat 37 g, Carbs 56 g, Added Sugar 5.6 g, Fiber 10.6 g

Sweet Potatoes with Asparagus, Eggplant and Halloumi

Ingredients

- 1 aubergine
- 9 tbsp. olive oil
- Chili flakes
- Salt
- Pepper
- 2 sweet potatoes
- 1 red chili pepper
- 2 tbsp. sunflower seeds
- 1 bunch green asparagus
- 4 tbsp. lemon juice
- 200 g chickpeas (can; drip weight)
- ½ bundle basil
- ½ bundle lemon balm
- 1 tsp. mustard
- ½ tsp. turmeric powder
- 1 tsp. honey
- 300 g halloumi

Directions: 45 min

1. Clean, wash and slice the eggplant. Heat 2 tablespoons of oil in a pan and sauté the aubergine slices in medium heat on both sides for 5–7 minutes until golden brown and season with chili flakes, salt, and pepper. Remove from the pan and set aside.

2. In the meantime, peel the sweet potato and cut it into cubes. Halve the chili lengthways, remove the stones, wash and cut into slices. Heat 1 tablespoon of oil in the pan, fry the sweet potato cubes in it for 10 minutes. Add 1 tbsp. sunflower seeds and chili slices and season with salt and pepper. Set aside as well.
3. Wash asparagus on the side, cut off the woody ends, peel the lower third of the stalks if necessary. Heat 1 tablespoon of oil in the pan, fry the asparagus in it for 5 minutes over medium heat. Deglaze with 1 tablespoon of lemon juice, pour in 2 tablespoons of water and cover and cook for another 3 minutes.
4. Rinse the chickpeas and let them drain. Wash the basil and lemon balm, shake dry and chop. Mix chickpeas with half of the herbs and 1 tablespoon of oil and season with salt and pepper.
5. Whisk the remaining oil with the rest of lemon juice, mustard, turmeric and honey, season with salt and pepper, and mix in the remaining herbs.
6. Cut the halloumi and slice in a hot pan on both sides for 5 minutes over medium heat until golden yellow.
7. Arrange sweet potatoes, aubergine slices on plates, serve with chickpeas, asparagus, and halloumi and drizzle with the dressing. Sprinkle with the remaining sunflower seeds.

Nutritional values | Calories 789

Protein 32 g, Fat 50 g, Carbs 53 g, Added Sugar 1 g, Fiber 10 g

CHAPTER SEVEN

Sirt Food Dinner Recipes

Pulled Chicken in Salad Tacos

Ingredients

- 300 g cooked chicken breast fillet
- 1 carrot
- 1 tomato
- 1 red onion
- Salt
- Pepper
- Chili flakes
- 200 g Greek yogurt
- 1 tsp. lemon juice
- 2 sticks celery
- 8 radicchio leaves
- 2 stems parsley

Directions: 20 min

1. Pluck chicken breast fillets with a fork. Clean, wash, and grate the carrot. Clean, wash, and cut the tomato into small cubes. Peel the onion, cut in half, and cut into fine strips. Mix everything together and season with salt, pepper, and chili flakes.

2. Mix the yogurt with lemon juice and season with salt, pepper, and chili flakes. Clean, wash, and chop the celery. Wash radicchio leaves and parsley and shake dry; Roughly chop parsley.
3. Spread the chicken mix into the lettuce leaves, pour celery and parsley over it and drizzle with yogurt sauce.

Nutritional values | Calories 198

Protein 22 g, Fat 9 g, Carbs 6 g, Fiber 2.6 g

Veal cabbage Rolls – Smarter with Capers, Garlic and Caraway Seeds

Ingredients
- 1 kg white cabbage (1 white cabbage)
- Salt
- 2nd onions
- 1 clove of garlic
- 3 tbsp. oil
- 700 g veal mince (order from the butcher)
- 40 g capers (glass; drained weight)
- 2 eggs
- Pepper
- 1 tsp. caraway seed
- 1 tbsp. paprika powder (sweet)
- 400 ml veal stock
- 125 ml soy cream

Directions: 25 min

1. Wash the cabbage and remove the outer leaves. Cut out the stalk in a wedge shape. Place a large pot of salted water and bring it to a boil.
2. In the meantime, remove 16 leaves from the cabbage one after the other, add to the boiling water and cook for 3-4 minutes
3. Lift out, rinse under running cold water and drain. Place on a kitchen towel, cover with a second towel and pat dry
4. Cut out the hard, middle leaf ribs.
5. Peel and finely chop onions and garlic. Heat 1 tablespoon of oil. Braise the onions and garlic until translucent.
6. Let cool in a bowl. Add minced meat, capers, eggs, salt and pepper and mix everything into a meat batter.
7. Put two cabbage leaves together and put one serving of mince on each leaf. Roll up tightly and tie with kitchen thread.
8. Heat the remaining oil in a saucepan and brown the eight cabbage rolls in it from each side.
9. Add the caraway and paprika powder. Pour veal stock into the pot and bring to a boil. Cover and braise the cabbage rolls over medium heat for 35–40 minutes, turn in between. Stir the soy cream into the sauce and let it boil for another 5 minutes. Season with salt and pepper. Put the cabbage roulades on a plate and serve with brown rice or mashed potatoes.

Nutritional values | Calories 452

Protein 57 g, Fat 21 g, Carbs 6 g, Roughage 5 g

Fish Cakes with Potato Salad

Ingredients

- 800 g stuck potatoes
- 100 g sugar snap
- ⅓ cucumber
- 200 g cherry tomatoes
- 10 g dill (0.5 bunch)
- 1 red onion
- 4 tbsp. olive oil
- 2 tbsp. apple cider vinegar
- 1 tsp. honey
- Salt
- Pepper
- 400 g cod fillet
- 2 tbsp. lemon juice
- 1 clove of garlic
- 50 g herbs (parsley, chives)
- 1 egg
- 50 g whole grain bread crumbs

Directions: 70 min

1. Wash potatoes and cook in boiling water over medium heat for 20–25 minutes. Then pour off and let cool.

2. In the meantime, clean and wash the sugar snap peas and cook them in a sieve over the steam of the boiling potatoes for 3–5 minutes. Quench and cut into fine strips. Clean and wash the cucumber and cut it into slices. Wash and halve tomatoes. Wash the dill, shake it dry and chop it roughly.
3. Peel the onion and cut into fine strips. Mix 2 tablespoons of oil with vinegar, honey and onion strips, salt, and pepper. Peel and slice potatoes. Fold in the sugar pods, cucumber slices, tomatoes, and dill and mix with the vinaigrette.
4. Rinse cod fillet, pat dry, cut into small pieces, puree with lemon juice. Peel, chop, and add garlic. Wash herbs, shake dry and chop. Mix everything together with egg, crumbs, salt, and pepper. Shape 8 meatballs with wet hands. Heat the remaining oil in a large pan. Fry meatballs in it on both sides over medium heat for 10 minutes and serve with the potato salad.

Nutritional values | Calories 421

Protein 27 g, Fat 13 g, Carbs 48 g, Added Sugar 1 g, Fiber 6.2 g

Egg and Avocado Sandwich with Crab Salad

Ingredients

- 2 eggs
- 1 organic lime
- 10 g ginger
- 1 tsp. honey

- 50 ml rapeseed oil
- 200 g Greek yogurt
- Salt
- Cayenne pepper
- 200 g north sea crabs (ready to cook; pre-cooked)
- 1 avocado
- 100 g radicchio
- 1 box shiso cress
- 8 slices whole-grain sandwich bread

Directions: 25 min

1. Boil eggs hard in boiling water in 8–9 minutes. Remove, quench under running cold water and peel.
2. While the eggs are boiling, rinse the lime hot, pat dry, and rub the skin. Halve the lime and squeeze out the juice. Peel and finely grate the ginger.
3. Puree the lime zest, 1 tsp. lime juice, ginger, honey, oil, and yogurt. Season with salt and cayenne pepper. Fold in the crab.
4. Halve the avocado, remove the stones, lift the pulp from the skin and cut into fine slices. Drizzle with a little lime juice. Wash the radicchio, shake it dry and remove the leaves from the stalk. Slice the eggs. Cut the cress from the bed.
5. Toast slices of bread. Cover 4 slices with eggs, crab salad, avocado, radicchio, and cress and cover with one slice of bread each. Halve sandwiches and arrange on plates.

Nutritional values | Calories 458

Protein 20 g, Fat 29 g, Carbs 29 g, Fiber 7.3 g

Vegetarian Lasagna - Smarter with Seitan and Spinach

Ingredients

- 225 g spinach leaves (frozen)
- 300 g seitan
- 2nd small carrots
- 2 sticks celery
- 1 onion
- 1 clove of garlic
- 2 tbsp. olive oil
- Salt
- Pepper
- 850 g canned tomatoes
- 200 ml classic vegetable broth
- 1 tsp. fennel seeds
- 30 g parmesan (1 piece)
- Nutmeg
- 225 g ricotta
- 16 whole-grain lasagna sheets
- Butter for the mold
- 150 g mozzarella

Directions: 40 min

1. Let the spinach thaw. Chop the seitan finely or put it through the middle slice of the meat grinder.
2. Wash and peel carrots. Wash, clean, remove and finely dice the celery. Peel and chop the onion and garlic.
3. Heat the oil in a saucepan and braise the carrots, celery, onions and garlic for 3 minutes over medium heat. Add since then and braise for 3 minutes while stirring. Season with salt and pepper.
4. Put the canned tomatoes and broth in the saucepan and cover and cook over medium heat for 20 minutes, stirring occasionally. Crush the fennel seeds, add and season the sauce with salt and pepper.
5. Meanwhile, finely grate the Parmesan. Rub off some nutmeg. Squeeze the spinach vigorously, roughly chop and stir in a bowl with the ricotta, parmesan, salt, pepper and nutmeg.
6. Lightly grease a baking dish (approx. 30 x 20 cm). Cover the bottom of the mold with a little sauce and smooth it out. Place 4 sheets of lasagna next to each other, if necessary, cut to size.
7. Add 1/3 of the spinach mixture and smooth out. Spread 1/4 of the sauce on top. Layer 4 lasagna sheets, 1/3 spinach and 1/4 sauce again, repeat the process again.
8. Place the last lasagna sheets on top and spread the rest of the sauce over them.

9. Drain the mozzarella and tear it into large pieces. Spread on the lasagna. Bake vegetarian lasagna in a preheated oven at 180 ° C (fan oven: 160 ° C, gas: levels 2–3) on the middle shelf for 35–40 minutes. Let the vegetarian lasagna rest for about 5 minutes before serving.

Nutritional values | Calories 633

Protein 50 g, Fat 25 g, Carbs 48 g, Fiber 7.5 g

Spaghetti with Salmon in Lemon Sauce

Ingredients

- 150 g salmon fillet (without skin)
- 100 g leek (1 thin stick)
- 100 g small carrots (2 small carrots)
- ½ organic lemon
- 2 stems parsley
- 150 g whole grain spaghetti
- Salt
- 2 tbsp. olive oil
- Pepper
- 100 ml of fish stock
- 150 ml of soy cream

Directions: 30 min

1. Wash salmon, pat dry and cut into 2 cm cubes.

2. Clean the leek, wash it, and cut it into thin rings. Peel the carrots and cut them into thin strips.
3. In the meantime, rinse the lemon half hot and rub dry. Peel the lemon peel thinly and cut into very fine strips. Squeeze lemon juice. Wash parsley, shake dry, pluck leaves and chop finely.
4. Cook the pasta bite-proof in saltwater according to the package instructions.
5. Heat oil in a pan. Season the salmon with pepper and fry all over in the hot oil for 3-4 minutes.
6. Take out the salmon, braise the leek rings, and carrot strips in the pan over medium heat for 3-4 minutes.
7. Take out the salmon, braise the leek rings, and carrot strips in the pan over medium heat for 3-4 minutes.
8. If necessary, salt the salmon, put it back in the pan, and heat briefly. Mix in the parsley.
9. Drain the pasta in a sieve and mix gently with the sauce. Season with salt and pepper and serve the spaghetti with salmon immediately.

Rice Shakshuka with Olives

Ingredients
- 120 g brown rice
- Salt
- 1 red onion

- 1 clove of garlic
- 2 tbsp. olive oil
- 100 g spinach (2 handfuls)
- 500 g roughly passed tomato
- ¼ tsp. chili flakes
- ½ tsp. cumin
- 2 tbsp. lemon juice
- Pepper
- 85 g black olives
- 4 eggs
- 100 g cherry tomatoes
- 10 g parsley (0.5 bunch)

Directions: 45 min

1. Cook rice covered with 250 ml of boiling salted water over low heat for 25–30 minutes. Remove from the stove, loosen and let swell for 5 minutes.
2. In the meantime, peel and finely chop the onion and garlic. Heat oil in a pan. Braise the onion and garlic for 2 minutes over medium heat. Wash the spinach, shake it dry, add it to the pan and let it collapse. Add the passed tomatoes and cook on a low heat for 5 minutes.
3. Add the cooked rice, stir in and season with chili flakes, cumin, lemon juice, salt, and pepper. Drain the olives and add them to the pan.

4. Smooth out the tomato mixture, form 4 small hollows and beat in 1 egg each. Cover the pan and cook over low heat for 20 minutes until the eggs are cooked.
5. In the meantime, wash, clean and halve cherry tomatoes. Add to the pan 5 minutes before the end of the cooking time. Wash the parsley, shake dry, roughly chop the leaves and garnish Shakshuka with it.

Nutritional values | Calories 367

Protein 16 g, Fat 15 g, Carbs 42 g, Fiber 7 g (23%)

Low Carb Pancakes with Spinach and Cheese

Ingredients

- 3 eggs
- 125 g low-fat quark
- 1 tsp. baking powder
- 100 g ground almond kernels
- 150 ml sparkling water
- 500 g spinach
- 1 onion
- 1 clove of garlic
- 3 tbsp. olive oil
- 250 g ricotta
- pepper
- 200 g mountain cheese (45% fat in dry matter)

- ½ chili pepper
- 30 g almond pencils (2 tbsp.)

Directions: 60 min

1. Separate eggs. Beat the egg whites until stiff, mix the egg yolks with the quark, baking powder and one pinch of salt, stir in the ground almonds. Carefully fold in the water and egg whites and let the dough soak for 10 minutes.
2. In the meantime, clean, wash and cut the spinach into strips. Peel onion and garlic and chop finely. Heat 1 tablespoon of oil in a pan. Braise both for 2 minutes over medium heat. Add spinach and sauté for 5 minutes. Then remove from the stove. Mix ricotta with spinach, season with salt and pepper.
3. Bake about 24 pancakes from the dough in the remaining oil in portions in about 5 minutes each. Place the pancakes on a baking sheet lined with baking paper, add the spinach mixture. Grate the cheese and sprinkle over the pancakes.
4. Bake pancakes in a preheated oven at 180 ° C (fan oven 160 ° C; gas: levels 2–3) for 15 minutes. Wash the chili, halve lengthways, core and chop. Garnish the pancakes with the almonds.

Baru Nut Bowl

Ingredients

- 300 g red cabbage
- Salt

- ¼ cucumber
- 2 tomatoes
- 2 zucchini
- ½ romaine lettuce
- 15 Baru nuts
- ½ lemon
- 4 tbsp. olive oil
- ½ tsp. mustard
- 1 tsp. honey
- Pepper
- 2 stems parsley

Directions: 30 min

1. Clean the red cabbage, remove the stem, wash and cut the remaining cabbage into fine strips. Place in a bowl, salt, and knead vigorously for 3 minutes.
2. Clean and wash the cucumber, tomato, and zucchini. Halve the cucumber lengthways and cut into fine slices, dice the tomatoes, and turn the zucchini into pasta with a spiral cutter. Clean, wash, and cut Romana into strips.
3. Squeeze the lemon and stir the juice together with three tablespoons of oil, mustard, honey, salt, and pepper into a dressing. Finely chop the Baru nuts and add about half to the dressing. Wash the parsley, shake dry and also add to the salad dressing.

4. Fry the zucchini in a pan with the remaining oil for about 5 minutes over medium heat. Season with salt and pepper.
5. Arrange the salad and vegetables in bowls, sprinkle with the remaining Baru nuts and drizzle the dressing over them.

Nutritional values | Calories 188

Protein 6 g, Fat 13 g, Carbs 12 g, Added Sugar 0.9 g, Fiber 5.7 g

Grilled Rosemary Sea Bream

Ingredients
- 500 g sea bream (shingled, ready to cook, a 2 doraden)
- Salt
- Pepper
- 1 clove of garlic
- 2 shallots
- 4 branches rosemary
- 5 tbsp. olive oil
- 1 small organic lemon
- 80 purslane (1 bunch)
- 10 g chervil (0.5 bunch)
- 10 g dill (0.5 bunch)
- 10 g parsley (0.5 bunch)
- 1 apple
- ½ bundle radish
- 40 g sprouts (e.g., radish or radish sprouts)

- 3 tbsp. balsamic vinegar
- 1 tsp. mustard
- 1 tsp. honey
- Edible violet flower

Directions: 30 min

1. Rinse the sea bream inside and outside and pat dry. Salt and pepper from all sides.
2. For the filling, peel garlic and shallots and finely dice both. Wash rosemary, shake dry. Put the garlic, shallots, and rosemary in the abdominal cavity of the sea bream and close the openings with 2 toothpicks each. Brush sea bream with 1 tablespoon of oil and place in grill baskets.
3. Rinse the lemon hot, pat dry, cut into thin slices and spread over the sea bream. Close the grill basket, place on a hot grill rack, and grill for 15-20 minutes, turning once.
4. In the meantime, wash purslane, chervil, dill, and parsley and shake dry. Pluck the leaves of chervil, dill, and parsley, mix with purslane and arrange on 4 plates.
5. Clean, wash, core the apple, and cut it into pencils. Clean, wash, and slice radishes. Rinse off the sprouts and drain well. Spread the apple, radishes, and sprouts on the salad.
6. Mix the balsamic vinegar with the mustard, honey, and remaining oil. Season with salt and pepper, pour over the salad

and garnish with violet flowers. Arrange sea bream with the salad.

Poached Eggs on Spinach with Red Wine Sauce

Ingredients
- 1 clove of garlic
- 3 tsp. olive oil
- 1 tsp. sugar
- 200 ml red wine
- Salt
- Pepper
- 1 shallot
- 250 g young spinach leaves
- Nutmeg
- 2 tbsp. white wine vinegar
- 4 eggs
- 2 slices whole-grain toast

Directions: 20 min
1. Peel and finely chop the garlic and braise in 1 teaspoon of oil. Sprinkle sugar, add red wine and bring to the boil. Reduce to 1/3 over medium heat. Salt, pepper, and keep warm.
2. Peel the shallot and dice finely. Wash the spinach well and let it drain. Heat the remaining oil in a pan, sauté the shallot in a

glassy heat. Add the spinach and let it collapse in 3-4 minutes. Add a little nutmeg, salt, and pepper.
3. Boil 1 liter of water with the vinegar. Carefully beat the eggs in a bowl so that the yolks remain intact.
4. Stir the boiling vinegar water vigorously with a whisk.
5. Now let the eggs slide in (by the rotation of the water they separate immediately).
6. Boil the water again. Then remove the saucepan from the hob and let the eggs cook (poach) for 3-4 minutes.
7. Lift out the eggs with a foam trowel and let them drain. Put the spinach on a plate and put the eggs on it. Drizzle with the wine sauce and serve. Toast the bread in the toaster and serve with it.

Nutritional values | Calories 358

Protein 21 g, Fat 19 g, Carbs 20 g, Added sugar 3 g, Fiber 6 g

Pasta with Minced Lamb Balls and Eggplant, Tomatoes and Sultanas

Ingredients

- 250 g lean minced lamb
- 2 tbsp. low-fat quark
- 1 egg
- 2 tbsp. breadcrumbs
- Salt
- Pepper

- 1 msp. cinnamon
- 200 g small eggplant (1 small eggplant)
- 1 onion
- 1 clove of garlic
- 2 tbsp. olive oil
- 150 g orecchiette pasta
- 2 tbsp. sultanas
- 400 g pizza tomatoes (can)
- 1 bay leaf
- 125 ml classic vegetable broth

Directions: 40 min

1. Mix minced lamb, quark, egg and breadcrumbs in a bowl. Season with salt, pepper and cinnamon.
2. Using wet hands, turn the chop into balls the size of a cherry. Chill briefly.
3. Clean, wash, dry the eggplant and cut into 5 mm cubes. Peel onion and garlic and chop finely.
4. Heat 1 tablespoon of oil in a pan and fry the meatballs in it until golden brown. Remove and set aside.
5. Wipe out the pan, then heat up the remaining oil. Add the eggplant cubes, onion and garlic and braise for 4-5 minutes, stirring. In the meantime, cook the pasta bite-proof in plenty of boiling salted water according to the package instructions.

6. Add the sultanas, tomatoes and bay leave to the pan. Pour in the broth and bring to the boil.
7. Cover and cook for 4 minutes over medium heat. Then put the meatballs in the pan and cook covered for another 5 minutes. Season with salt and pepper and keep warm.
8. Drain the pasta in a sieve and let it drain. Serve mixed with the sauce.

Nutritional values | Calories 724

Protein 53 g, Fat 22 g, Carbs 75 g, Fiber 8.5 g

CHAPTER EIGHT

Dessert Recipes

Blackberry Quark Tartlets

Ingredients

- 250 g whole meal spelled flour + 1 heaped tablespoon for processing
- 35 g cocoa powder (5 tbsp.; heavily oiled)
- 1 pinch salt
- 50 g raw cane sugar
- 125 g cold, diced butter
- 2 eggs
- 250 g blackberry
- 1 organic lime
- 1 vanilla bean
- 100 g creme fraiche cheese
- 250 g low-fat quark
- 1 tbsp. maple syrup
- 15 g dark chocolate at will (at least 70% cocoa)

Directions: 50 min

1. Mix the flour with the cocoa powder, salt, and sugar. Add butter and one egg and use your hands to make a smooth dough. Cover and chill for 30 minutes.

2. In the meantime, carefully wash the blackberries, drain, and cut them in half or quarter, depending on the size. Rinse lime hot, rub dry, rub the peel; Halve the lime and squeeze out the juice. Halve the vanilla pod lengthways and scrape out the pulp. Mix the crème fraîche with the quark, half of the lime peel, 1–2 sprinkles of lime juice, maple syrup, vanilla pulp, and the rest of the egg.
3. Divide the dough into 4 equal pieces, roll them out on a floured work surface (approx. 14 cm each) and put them in 4 tartlet molds.
4. Spread the curd mixture on the molds and pour the blackberries over it. Bake the tartlets in a preheated oven at 180 ° C (fan oven 160 ° C; gas: levels 2–3) for 30–35 minutes.
5. Remove the tartlets from the oven, let them cool for 10 minutes and remove them from the molds. Garnish the blackberry quark tartlets with the remaining lime peel and grated chocolate as desired.

Nutritional values | Calories 824

Protein 16 g, Fat 51 g, Carbs 76 g, Added Sugar 24 g, Roughage 4 g

Chocolate Strawberries with Cardamom

Ingredients

- 400 g strawberries
- 2 cardamom capsules

- 100 g dark chocolate coating (at least 72% cocoa)

Directions: 20 min

1. Place the strawberries in a sieve, wash carefully and pat dry.
2. Break open cardamom capsules and remove seeds. Finely crush cardamom seeds in a mortar.
3. Roughly chop the couverture and put it in a small beater. Add cardamom.
4. Allow the chocolate to melt in a hot water bath while stirring.
5. Hold the strawberries by the stem and dip 2/3 in succession in the liquid chocolate coating.
6. Place the chocolate strawberries on baking paper and let the couverture dry. Chill chocolate strawberries until ready to serve.

Nutritional values | Calories 130

Protein 3 g, Fat 5 g, Carbs 17 g, Added Sugar 6 g, Roughage 6.5 g

Chocolate Fruit Cake

Ingredients

- 300 g prunes
- 300 g dried fig
- 200 g baked fruit
- 200 g almond kernels
- 150 g hazelnuts
- 5 eggs
- 125 g butter

- 1 tbsp. honey
- 200 g spelled flour
- 1 pinch ground carnation
- ½ tsp. ground ginger
- 1 tbsp. cinnamon
- 100 g dark chocolate
- 20 g coconut oil

Directions: 45 min

1. Roughly chop plums, figs and baked fruit. Chop nuts with a knife or briefly put them in a Blitzhacker. Separate eggs beat egg whites with a hand mixer to firm snow. Whisk the butter and honey until fluffy, then add the egg yolk and flour and stir to a smooth dough. Knead the fruits, nuts and spices under the dough and carefully fold in the egg whites.
2. Line a baking tin with baking paper and pour in the dough. Bake in a preheated oven at 175 ° C (fan oven: 150 ° C; gas: speed 2) for about 60 minutes.
3. Take the cake out of the oven and let it cool. Meanwhile, chop the chocolate and melt together with coconut oil over a hot water bath. Warp the cake with the chocolate.

Pear Chocolate Cake with Pistachios

Ingredients

For the dough

- 1 can pear half approx. 825 ml content
- 100 g dark chocolate
- 200 g butter
- 200 g sugar
- 4 eggs
- Salt
- 1 splash lemon juice
- 50 ml of milk
- 350 g flour
- 2 tsp. baking powder
- 3 tbsp. cocoa powder
- Butter for the mold
- Flour for the mold
- 125 g dark chocolate coating
- 2 tbsp. pistachio nuts unsalted

Directions: 40 min

1. Preheat the oven to 180 ° C top and bottom heat. Butter a springform pan and dust with flour. Drain the pears. Chop the chocolate. Mix the butter with the sugar until frothy. Gradually stir in the eggs and a pinch of salt. Then add the milk and lemon juice. Mix the flour with the baking powder and the cocoa and sieve onto the egg and butter mixture. Fold in the chopped chocolate. Put half of the dough in the mold. Cut the pear halves into pieces and spread them on the dough, put the rest of the

dough on top, and spread evenly. Bake the cake in the preheated oven for about 45 minutes. (Make a stick test!) Take out the finished cake, let it cool briefly, carefully remove it from the mold and let it cool.

2. Chop the dark chocolate coating, melt it, temper it and pour it over the cake. Sprinkle with the pistachios.

Nut and Chocolate Cookies

Ingredients

- 150 g ground hazelnuts
- 200 g soft butter
- 100 g powdered sugar
- 2 egg yolks
- 1 pinch mace
- 1 pinch cinnamon
- 150 g flour
- 4 tbsp. nut nougat cream more if needed
- 100 g dark chocolate coating
- 1 tsp. oil
- Flour for the work surface

Directions: 1 h 45 min

1. Roast the nuts briefly / lightly in a pan without fat. Beat butter, powdered sugar, egg yolks, and spices until fluffy. Knead quickly

with the flour and the cooled nuts to a smooth dough. Wrap in cling film and chill for 30 minutes.
2. Preheat the oven. Cover the baking sheets with baking paper. Roll out the dough 3-4 mm thin on a floured surface. Cut out circles (approx. 3 cm Ø) and place them on the sheet. Bake in the oven at 175 ° C (middle, fan oven 160 ° C) for 10 minutes. Pull off the sheet with the paper and let cool.
3. Spread a thin layer of nut nougat cream on half of the cookies and put the other half together. Melt the couverture with oil in a water bath and immerse the cookies on both sides so that there is a free gap in the middle. Let it dry on sandwich paper and pack it in a transparent film to give away.

Nutritional values | Calories 75

Cauliflower and Cheese Gressinos

Ingredients
- 200g Cauliflower
- 50g mozzarella cheese
- 4 eggs
- 2 cc ground oregano
- Salt & pepper

Directions: 20 min
1. Put the gratin paper into the baking dish. Make sure the cauliflower is cut into a rapier.

2. Add rapier to the food processor and beat it until the cauliflower looks like rice.
3. Place the cauliflower in a microwave and place it in a special container with a lid for 2-3 minutes.
4. Remove cauliflower from the microwave and pour it into a large bowl. Add 4 eggs, 40g mozzarella cheese, oregano, garlic, salt, and pepper.
5. Made all the dough and made a dough.
6. Bake only the crust at medium temperature for about 25 minutes or until golden (not topped yet).
7. Once baked, spread the remaining mozzarella cheese and bake for another 5 minutes or until the cheese melts.
8. Served by cutting into bread crumbs.

Apple Cheesecake

Ingredients

For the very low carb apple cheesecake (approx. 12 pieces) you need:
- 250g mascarpone
- 250g low-fat curd
- 3 eggs
- 1 pack of custard
- 1 teaspoon Baking powder
- 1 packet of vanilla sugar
- 1 pinch of salt

- 2-3 apples

Directions: 20 min

1. Peel, quarter and scratch the apples.
2. Mix the remaining ingredients together. First, the dry ones, then the others.
3. Line 28cm springform pan with baking paper. Fill in the dough. Press in the apples.
4. 45 minutes at 160 ° C (circulation) baking, allow cooling thereafter in the oven slightly open.
5. Preparation time without baking 10-15 minutes. Depending on how quickly you can peel apples.

Halloween Desserts

Ingredients

- 210 g wholemeal spelled flour
- 7 tbsp. whole cane sugar
- 40 g butter
- 280 g dark chocolate (at least 70% cocoa)
- 1 l milk (3.5% fat)
- 1 vanilla bean
- 1 map. orange peel
- 6 egg yolks
- 2 tbsp. food starch
- 5 tsp. cocoa powder

Directions: 60 min

1. For the biscuits, quickly knead 50 g of flour, 1 tbsp. of whole cane sugar, and 25 g of butter into a smooth shortcrust pastry. Chill wrapped in cling film for about 30 minutes.
2. Roughly chop chocolate. Remove ten tablespoons from the milk and set aside. Slice the vanilla pod lengthways and scrape out the pulp.
3. Put the remaining milk in a saucepan with 240 g chocolate, vanilla pulp, pod, and orange zest and heat while stirring. In a bowl, stir the egg yolks with 4 tablespoons of whole cane sugar until frothy, stir in the cornstarch and the milk set aside.
4. Stir in the hot chocolate milk and bring back to the boil in the pot and let it thicken. Remove the mixture from the stove, continue beating briefly, remove the vanilla pod, and let the pudding cool in glasses.
5. Sprinkle biscuit dough with 1 tsp. cocoa powder and work in as a marbling. Roll out the dough and cut out tombstones with an oval cutter. Place cookies on a baking sheet covered with baking paper and bake for 10–15 minutes in a preheated oven at 180 ° C (forced air 160 ° C; gas: setting 2-3). Take out and let cool.
6. To decorate, mix the remaining ingredients (cocoa powder, flour, butter, whole cane sugar) and crumble. Melt the last 40 g of dark chocolate, fill it into a small piping bag and label it with the cookies.

7. Sprinkle the pudding with the chocolate crumbs, insert the labeled cookies as a 'gravestone' and decorate them with flowers as you like

Nutritional values | Calories 506

Protein 14 g, Fat 23 g, Carbs 60 g, Added Sugar 27 g, Fiber 6 g

Zucchini and Cheese Quiche

Ingredients

- Coconut oil, butter or ghee for greasing
- 2-3 large zucchini
- 1 tsp. salt
- 550 g ricotta
- 75 g grated parmesan
- Chopped 2 spring onions
- 2 cloves of garlic chopped
- 4 tbsp. chopped dill
- 1 lemon zest
- 2 large eggs beaten
- 100-200 g crumbled feta, depending on your taste

Directions: 25 min

1. Preheat the oven to 170 ° C circulating air (190 ° C without circulating air) and grease a medium-sized baking dish or a springform pan with a diameter of 23 cm.

2. Grate the zucchini in the food processor. Mix with salt put in a colander or coarse sieve and let stand for 15 minutes.
3. Then squeeze out the liquid thoroughly with your fingers or a spoon.
4. Mix ricotta, parmesan, spring onions, garlic, dill and lemon zest.
5. Add the whisked eggs and zucchini.
6. Fill in the baking pan and bake for an hour. Spread the feta on top and continue baking until the cheese has melted.

Nutritional values | Calories 332

Protein 12 g, Fat 23 g, Carbs 20 g, Added Sugar 3 g, Fiber 2 g

Asparagus Quiche

Ingredients
- 1 pack of ready-made shortcrust pastry
- 1 bunch of green asparagus
- 75 g ham cubes
- 3 eggs
- 75 g gouda
- 200 ml cream
- Salt, pepper, nutmeg

Directions: 15 min
1. Grease the mold, flour and lay the shortcrust pastry in it. Pick in a couple of places with a fork so that the ground does not make waves.

2. Preheat the oven to 190 ° C and bake the base blind for 15 minutes. Let cool down.
3. Dice the cheese and cut the asparagus into bite-size pieces. Spread both together with the ham cubes on the dough. As a decoration, you can pick up a few asparagus spears and put them on top at the end - very pretty :-)
4. Whisk the eggs and cream and season with salt, pepper, and nutmeg. Then pour over the other ingredients.
5. The whole now about 30-40 minutes to bake, is hammered to the egg.

Nutritional values | Calories: 397

Protein: 15 g, Carbs 13 g, Fat 30 g, Added Sugar 2 g, Fiber 1 g

Spelled Waffles with Cherry Sauce

Ingredients

- 800 g cherries (fresh or from the glass)
- 1 organic lemon
- 1 tbsp. food starch
- 75 g sugar
- Ground cinnamon
- 150 ml dry red wine
- 250 ml cherry juice
- 3 eggs
- 250 g whole meal spelled flour

- 1 map. baking powder
- 250 g buttermilk
- 3 tbsp. sunflower oil
- Salt
- 1 tsp. yogurt butter
- Powdered sugar as desired

Directions: 60 min

1. Wash and stone fresh cherries for the compote; Drain the cherries from the glass.
2. Wash the lemon with hot water, grate the peel, squeeze out the juice and stir the starch with water. Bring 25 g sugar with cinnamon, lemon juice, red wine, and cherry juice to a boil in a saucepan. Simmer on medium heat for about 5 minutes. Stir in cornstarch and let the sauce thicken. Add the cherries and lemon zest and let cool in a bowl.
3. Separate the eggs for the waffles and stir the egg yolks with the remaining sugar until creamy. Mix the flour with the baking powder, alternating with buttermilk and oil, stir into the egg yolk mixture. Beat the egg whites with a pinch of salt and fold in.
4. Heat up the waffle iron. Grease with yogurt butter, add 2 tablespoons of batter. Bake golden brown waffles one after the other. Dust the waffles with icing sugar and serve with cherries.

Nutritional values | Calories 437

Protein 11 g, Fat 11 g, Carbs 68 g, Added Sugar 13 g, Fiber 6 g

Strawberry Cream Cake

Ingredients

- 260 g wholemeal emmer or spelled flour
- 150 g coconut sugar
- 1 egg
- 150 g cold butter
- 5 sheets white gelatin
- 600 g strawberries
- 1 tbsp. lemon juice
- 250 g mascarpone
- 250 g low-fat quark
- 200 ml whipped cream
- 20 g cornstarch (1 tbsp.)
- 250 ml strawberry juice or red currant juice

Directions: 1 h 45 min

1. Mix 250 g of flour with 50 g of coconut blossom sugar and pile on the worktop. Push a well in by hand, whisk in the egg, and spread butter in pieces around the dough. Chop all ingredients with a knife so that crumbs are formed, quickly process into a dough, and put in cling film and put in the fridge for 30 minutes.
2. Line the bottom of a springform pan with baking paper. Roll out the dough slightly larger than the springform tin (26 cm in diameter) on a work surface dusted with remaining flour and line the bottom with it. Bake in a preheated oven at 200 ° C (fan

oven 180 °C; gas: setting 3) for about 15–20 minutes, remove and let cool.

3. Soak the gelatine in cold water. Clean and wash strawberries, cut 1/3 of them into slices, halve 1/3 and puree 1/3 with 50 g coconut blossom sugar and lemon juice. Push the puree through a hair strainer and mix with the mascarpone, low-fat curd cheese, and the remaining coconut blossom sugar. Whip the cream until stiff and carefully fold into the strawberry mascarpone mixture. Heat the soaked gelatin in a saucepan and add 2 tablespoons of the mascarpone cream, stir and stir into the rest of the mascarpone cream.

4. Line up the strawberry slices on the edge of the springform pan (set aside 8 slices for decoration) and place the strawberry halves on the ground with the cut side down. Spread the mascarpone cream over the strawberries, smooth out and leave in the fridge for 1-2 hours.

5. Arrange the remaining strawberry slices in a circle in the middle of the cake. Mix the starch with a little strawberry juice. Bring the rest of the strawberry juice to a boil in a saucepan and stir in the starch mixture. Let it cook for about 1/2 minute, then pour the icing evenly onto the cake. Chill again for at least 1 hour. Remove from the springform pan and serve cut into pieces.

Brownie Cheesecake

Ingredients

- 100 g dark chocolate (70% cocoa content)
- 125 g room temperature butter
- 100 g raw cane sugar
- 3 eggs
- 300 g quark (20% fat)
- 125 g spelled flour type 1050
- ½ packet baking powder
- ½ tsp. vanilla powder
- 1 pinch salt

Directions: 30 min

1. For the chocolate mass, roughly chop the chocolate over a hot, non-boiling water bath. Then let it cool down a bit.
2. In a bowl, stir the butter with the raw cane sugar until creamy. Mix in the eggs and the curd. Mix the flour with the baking powder, vanilla, and salt and stir the flour mixture into the dough. Divide the dough and stir in the chocolate under half.
3. Fill the dough alternately in 3-4 layers in the baking mold and carefully marble with a fork. Bake in a preheated oven at 180 ° C (fan oven 160 ° C; gas: levels 2–3) for 30 minutes. Take out and let cool on a wire rack. Cut into pieces to serve.

Nutritional values | Calories 139

Protein 4 g, Fat 8 g, Carbs 12 g, Added Sugar 7 g, Fiber 1 g

CHAPTER NINE

Snacks Recipes

Pancake Skewers with Fruits

Ingredients

- 100 g dark chocolate (at least 70% cocoa)
- 3 bananas
- 200 g buttermilk
- 100 g yogurt (3.5% fat)
- 3 eggs
- 3 tbsp. rapeseed oil
- 100 g 5-grain flakes (or oat flakes)
- 100 g wholemeal spelled flour
- ½ tsp. baking soda
- 1 tsp. baking powder
- 1 tbsp. whole cane sugar
- 150 g strawberries
- 2 handfuls blueberries

Directions: 30 min

1. Chop the chocolate and melt over a warm water bath. Peel bananas, slice them, pull them through the chocolate and let them dry on baking paper.

2. Mix the buttermilk, yogurt, eggs, and 1 tablespoon of rapeseed oil. Chop the 5-grain flakes very finely in a blender, then add the flour, baking soda, baking powder, and sugar to another bowl and mix. Add liquid ingredients and make a smooth dough. Let it rest for about 10 minutes.
3. In the meantime, clean, wash, and slice strawberries. Wash and pat the blueberries dry.
4. Heat some oil in a pan and bake small pancakes with 1 teaspoon of dough each over medium heat.
5. Serve the skewers, skewer a blueberry, then a strawberry slice, a pancake, a chocolate banana, and a pancake.

Tomato and Zucchini Salad with Feta

Ingredients

- 2 zucchini
- 4 tbsp. olive oil
- Salt
- Pepper
- 400 g tomatoes
- 200 g cherry tomatoes
- 3 spring onions
- 1 bunch basil (20 g)
- 2 tbsp. apple cider vinegar
- 100 g feta (45% fat in dry matter)

Directions: 10 min

1. Clean, wash, and cut zucchini. Heat 1 tablespoon of oil in a pan, fry the zucchini in it over medium heat for 5 minutes. Season with salt and pepper.
2. Clean, wash, and chop tomatoes. Wash and halve cherry tomatoes. Wash the spring onions and cut them into rings. Wash the basil, shake dry and pick the leaves.
3. Mix zucchini, tomatoes, cherry tomatoes, and basil. Add the remaining oil and apple cider vinegar, mix, and season with salt and pepper. Crumble the feta. Serve the salad sprinkled with feta.

Nutritional values | Calories 234

Protein 9 g, Fat 17 g, Carbs 10 g, Fiber 4.7 g

Blueberry and Coconut Rolls

Ingredients

- 150 g wholemeal flour
- 150 g spelled flour
- 1½ tsp. baking powder
- 1 pinch salt
- 50 g raw cane sugar
- 4 tbsp. rapeseed oil
- 250 g low-fat quark
- 1 egg

- 5 tbsp. milk (3.5% fat)
- 120 g blueberries
- 4 tbsp. grated coconut

Directions: 15 min

1. Put the flour with baking powder and salt in a bowl. Add sugar and mix. Add rapeseed oil, curd cheese, egg and 4 tablespoons of milk and use a hand mixer for kneading into a smooth dough.
2. Wash the blueberries, pat dry and fold in together with the grated coconut under the dough.
3. Line a baking sheet with parchment paper. Form 9 round rolls with floured hands and place them on the baking sheet. Brush the blueberry and coconut buns with the remaining milk and bake in a preheated oven at 200 ° C (fan oven 180 ° C; gas: setting 3) for 12–15 minutes.

Nutritional values | Calories 227

Protein 9 g, Fat 8 g, Carbs 28 g, Added Sugar 5.3 g, Fiber 4.2 g

Brain Food Cookies

Ingredients

- 150 g spelled flour type 1050
- 1 tsp. baking powder
- 100 g whole cane sugar
- 1 pinch salt
- 120 g room temperature butter

- 3 ripe bananas
- 1 egg
- 150 g pithy oatmeal
- 60 g donated almonds
- 1 tbsp. cocoa nibs
- 2 tbsp. chocolate drop (made from dark chocolate; 15 g)

Directions: 30 min

1. Mix the flour with the baking powder, sugar and 1 pinch of salt. Add the butter in pieces and mix. Peel the bananas, mash them with a fork and add them to the dough together with the egg and stir well with a hand mixer. Fold in the oatmeal, almonds, cocoa nibs and half of the chocolate drops.
2. Line a baking sheet with parchment paper. Place the dough on the baking sheet with a tablespoon, leaving enough space between the cookies. Sprinkle with the remaining chocolate drops and bake in a preheated oven at 200 ° C (fan oven 180 ° C; gas: setting 3) for 10–15 minutes. Then let it cool down on a wire rack.

Nutritional values | Calories 110

Protein 2 g, Fat 6 g, Carbs 12 g, Added Sugar 3.8 g, Fiber 1.1 g

Chocolate Granola Bars

Ingredients

- 50 g dried date (without stone)

- 100 g nut mix
- 20 g heavily deoiled cocoa powder
- 250 g pithy oatmeal
- 30 g linseed
- 200 g dark chocolate (at least 70% cocoa)
- 50 g hazelnut butter
- 300 g applesauce or marrow
- 20 g unsweetened spelled flakes
- 20 g puffed amaranth
- 2 tsp. cinnamon
- 1 map. vanilla powder
- ½ tsp. ground cardamom
- 1 tsp. organic orange peel

Directions: 15 min

1. Roughly chop dates, nuts, cocoa powder, oatmeal and flaxseed in a food processor or a powerful blender.
2. Chop 100 g chocolate and stir together with hazelnut butter, apple pulp, spelled flakes, amaranth, and spices into the date mix.
3. Pour the mixture into a baking tin lined with baking paper and press well. Bake in a preheated oven at 200 ° C (fan oven 180 ° C, gas: setting 3) for 20–25 minutes until the surface is browned. Then let cool in the mold.

4. Remove the chocolate muesli bar from the mold and the baking paper, cut into 16 bars. Melt the remaining chocolate, decorate the bar with it and let it cool.

Nutritional values | Calories 213

Protein 7 g, Fat 11 g, Carbs 22 g, Added Sugar 5.4 g, Fiber 5.1 g

Blueberry and Coconut Rolls

Ingredients

- 150 g wholemeal flour
- 150 g spelled flour
- 1½ tsp. baking powder
- 1 pinch salt
- 50 g raw cane sugar
- 4 tbsp. rapeseed oil
- 250 g low-fat quark
- 1 egg
- 5 tbsp. milk (3.5% fat)
- 120 g blueberries
- 4 tbsp. grated coconut

Directions: 45 min

1. Put the flour with baking powder and salt in a bowl. Add sugar and mix. Add rapeseed oil, quark, egg and 4 tablespoons of milk and use a hand mixer for kneading into a smooth dough.

2. Wash the blueberries, pat dry, and fold in together with the grated coconut under the dough.
3. Line a baking sheet with parchment paper. Form 9 round rolls with floured hands and place them on the baking sheet. Brush the blueberry and coconut rolls with the remaining milk and bake in a preheated oven at 200 ° C (fan oven 180 ° C; gas: setting 3) for 12–15 minutes.

Vanilla Energy Balls with Coconut Shell

Ingredients

- 100 g almond kernels
- 200 g dried date (pitted)
- ½ tsp. vanilla powder
- 20 g grated coconut (approx. 2 tbsp.)

Directions: 50 min

1. Put almonds, dates, and vanilla powder in a food processor or strong blender and chop into sticky mush.
2. Form balls of equal size from the mass.
3. Put coconut flakes on a flat plate. Roll the vanilla energy balls in the coconut flakes and press them down lightly.
4. Place the vanilla energy balls in an airtight sealable box and keep in the fridge.

Nutritional values | Calories 517

Protein 35 g, Fat 26 g, Carbs 32 g, Fiber 5.5 g

Wake-Up Energy Balls

Ingredients

- 120 g dried dates (without stone)
- 120 g walnuts
- 6 str. tel cocoa powder (18 g; heavily oiled)
- 1 pinch salt
- 1 pinch vanilla powder
- 2 tbsp. coffee bean (30 g)
- 2 tsp. ground coffee (10 g)
- Chili flakes

Directions: 15 min

1. Put the dates together with the nuts, 4 teaspoons of cocoa powder, 1 pinch of salt and vanilla powder in a blender and puree until you get a homogeneous dough.
2. Add the coffee beans and mix briefly. Cut 16 portions of dough with a tablespoon and shape into balls.
3. Mix the remaining cocoa powder with coffee powder and chili flakes and roll the energy balls in it.

Nutritional values | Calories 81

Protein 2 g, Fat 6 g, Carbs 6 g, Fiber 2.4 g

Brain Food Cookies

Ingredients

- 150 g spelled flour type 1050
- 1 tsp. baking powder
- 100 g whole cane sugar
- 1 pinch salt
- 120 g room temperature butter
- 3 ripe bananas
- 1 egg
- 150 g pithy oatmeal
- 60 g donated almonds
- 1 tbsp. cocoa nibs
- 2 tbsp. chocolate drop (made from dark chocolate; 15 g)

Directions: 30 min

1. Mix the flour with the baking powder, sugar and 1 pinch of salt. Add the butter in pieces and mix. Peel the bananas, mash them with a fork and add them to the dough together with the egg and stir well with a hand mixer. Fold in the oatmeal, almonds, cocoa nibs and half of the chocolate drops.
2. Line a baking sheet with parchment paper. Place the dough on the baking sheet with a tablespoon, leaving enough space between the cookies. Sprinkle with the remaining chocolate drops and bake in a preheated oven at 200 ° C (fan oven 180 °

C; gas: setting 3) for 10–15 minutes. Then let it cool down on a wire rack.

Chocolate Granola Bars

Ingredients

- 50 g dried date (without stone)
- 100 g nut mix
- 20 g heavily deoiled cocoa powder
- 250 g pithy oatmeal
- 30 g linseed
- 200 g dark chocolate (at least 70% cocoa)
- 50 g hazelnut butter
- 300 g applesauce or marrow
- 20 g unsweetened spelled flakes
- 20 g puffed amaranth
- 2 tsp. cinnamon
- 1 MSP. vanilla powder
- ½ tsp. ground cardamom
- 1 tsp. organic orange peel

Directions: 35 min

- Roughly chop dates, nuts, cocoa powder, oatmeal and flaxseed in a food processor or a powerful blender.
- Chop 100 g chocolate and stir into the date mix together with hazelnut butter, apple pulp, spelled flakes, amaranth and spices.

- Pour the mixture into a baking tin lined with baking paper and press well. Bake in a preheated oven at 200 ° C (fan oven 180 ° C, gas: setting 3) for 20–25 minutes until the surface is browned. Then let cool in the mold.
- Remove the chocolate muesli bar from the mold and the baking paper, cut into 16 bars. Melt the remaining chocolate, decorate the bar with it and let it cool.

Kale Avocado and Chili Dip with Keto Crackers

Ingredients

- 75 g linseed
- 75 g pumpkin seeds
- 50 g sesame seeds
- 40 g almond flour (4 fl. el)
- 2 tbsp. slightly liquid coconut oil
- Salt
- 3 avocados
- 200 g kale
- 1 small green chili pepper
- 4 tbsp. olive oil
- 1 organic lemon (zest and juice)
- Pepper
- 1 handful watercress (5 g)

Directions: 20 min

1. Place seeds and kernels with almond flour in a bowl, pour 125 ml of hot water over them and let them steep for 10 minutes. Then add coconut oil, season with salt, and mix everything.
2. Spread the mixture thinly on a baking sheet covered with baking paper, about 5 mm thin. Bake crackers in a preheated oven at 175 ° C top and bottom heat (gas: level 2–3) for 25–30 minutes. Then remove, let cool for 10 minutes and break the crackers into pieces.
3. In the meantime, halve the avocados for the dip, remove the stones, lift the pulp out of the bowl with a spoon, and roughly dice. Clean kale, pluck the green from the stems, wash, shake dry and cut into small pieces. Halve, chop, wash, and chop lengthways. Put the avocado, kale, and chili together with olive oil in a blender and coarsely puree. Season everything with salt, lemon peel and juice, and pepper. Fill the dip into a bowl, garnish with watercress and serve with the crackers.

Nutritional values | Calories 337

Protein 10 g, Fat 31 g, Carbs 6 g, Fiber 8.3 g

Grilled Eggplant Rolls with Walnut and Feta Filling

Ingredients

- 2nd eggplants
- Salt

- 100 g instant couscous
- 1 clove of garlic
- 1 bunch parsley (20 g)
- 5 dried dates
- 10 g ginger root
- 60 g walnut kernels
- 200 g feta (45% fat in dry matter)
- 3 tbsp. lemon juice
- ½ tsp. rose hot paprika powder
- Pepper
- 2 tbsp. olive oil

Directions: 30 min

1. Clean and wash the eggplants, cut lengthways into 1/2 cm slices, sprinkle with a little salt, and set aside for 10 minutes.
2. Pour 250 ml of boiling water over the couscous and let it soak for 10 minutes. In the meantime, peel the garlic, wash the parsley and shake dry, set aside 1 handful. Core the dates. Peel the ginger. Finely chop garlic, remaining parsley, dates, ginger, and walnuts.
3. Put the feta in a bowl and crumble finely. Add the couscous, 2 tablespoons of lemon juice, garlic, chopped parsley, dates, ginger, and chopped walnuts and knead everything well with your hands. Season with paprika powder and pepper.

4. Dab eggplant slices with a clean cloth and brush with olive oil on both sides.
5. Heat the grill pan. Fry the aubergine slices on each side for 1-2 minutes over medium heat, then let cool for about 5 minutes. Put some of the feta mass on each slice, roll it up and place it on a plate or plate with the end down and put the parsley aside.

Nutritional values | Calories 442

Protein 15 g, Fat 28 g, Carbs 31 g, Fiber 5.3 g

CHAPTER TEN

Fruit Juice

Drinks with Oranges

Ingredients

- 1/3 glass of orange juice (freshly squeezed or direct juice)
- Hot water
- Small piece of ginger
- Orange slice for decoration

Directions: 10 min

1. Peel the ginger and cut it into thin slices.
2. Then pour up to 2/3 in a large beaker with hot water and stir.
3. Add the orange juice.
4. Decorate with the orange slice

Orange and Mandarin Liqueur

Ingredients

- 2 large oranges
- 2 tangerines
- 1 small lemon
- 300 g white sugar candy
- 1 stick of vanilla
- 50 ml of orange juice

- 250 ml double grain

Directions: 5 min

1. Put the sugar candy in a bottle or in a screw-top jar.
2. Pour the citrus into small pieces and remove the skin.
3. Pour in the orange juice.
4. Add the vanilla stick.
5. Baste with the double grain and fill up to the top of the bottle if desired.
6. Close the bottle.
7. Shake daily until the sugar candy has dissolved.
8. After 2 - 3 weeks, pour the liqueur through a sieve and pour it back into the bottle.

Nutritional value | Calories: 242

Protein: 1 g, Fat: 1 g, Carbs: 27g, Added Sugar 26 g, Fiber 1 g

Pear and Lime Marmalade

Ingredients

- 3-4 untreated limes
- 1 kg ripe pears
- 500 g jam sugar 2: 1

Directions: 20 min

1. Wash 2 limes and grate dry.
2. Peel the peels thinly with the zest ripper.

3. Then cut all limes in half and squeeze them out. Measure out 100 ml of lime juice.
4. Wash and peel the pears, remove the core and then quarter them. Weigh 900 g of pulp.
5. Then puree the pears together with the lime juice.
6. Now put the pear puree together with the lime peels and the jellied sugar in a saucepan.
7. Bring all ingredients to the boil together.
8. Simmer for 4 minutes, stirring, taking care not to burn anything.
9. Make a gelation test with a small blob on a cold saucer. If this becomes solid in a short time, the jam is ready.
10. Remove any foam that may have formed with a trowel, but you can also simply stir it in.
11. Then pour the hot mass into hot rinsed jars, close and let stand upside down

Nutritional value | Calories 130

Carbs 33 g, Added Sugar 31 g, Fiber 1 g

Kiwi Yogurt Ice Cream

Ingredients

- 360 ml of yogurt
- 8 kiwi fruits
- 150 g of sugar
- 45 ml of lemon juice

- 30 ml orange liqueur (e.g., Grand Marnier)

Directions: 3 h 2 min

1. Peel and roughly cut the kiwi into cubes.
2. Mix the kiwi cubes and sugar and let them steep for 20 minutes.
3. Then puree all ingredients in the blender.
4. Place in the ice maker and allow to freeze according to the instructions.

Christmas Cocktail - Vegan Eggnog

Ingredients

- 1 cup cashew nuts
- 1 cup soy or almond milk
- 2-3 glasses of water
- About 5 pieces of dates (more if you like sweeter drinks)
- 2-3 scoops of brandy or whiskey
- 1 tablespoon lemon juice (optional, to taste)
- 1-2 teaspoons cinnamon
- ½ teaspoons ground anise
- ½ teaspoons ground ginger
- 2 pinches nutmeg
- Pinch of salt

Directions: 10 min

1. Pour dates and cashews with boiling water and leave to soak for 20 minutes. Transfer the remaining ingredients to the blender dish and finally add the drained nuts and dates.
2. Mix thoroughly in a high-speed blender for a few minutes, until a thick and creamy cocktail without lumps is formed. If your blender can't do it, mix the cashews with water first and strain them with gauze.
3. Season the cocktail with more lemon juice and salt to taste, and if you prefer sweeter drinks, add 2-3 pieces of dates. Serve it chilled with a pinch of cinnamon

Watercress Smoothie

Ingredients

- 150 g watercress
- 1 small onion
- ½ cucumber
- 1 tbsp. lemon juice
- 200 ml mineral water
- Salt
- Pepper
- 4 tbsp. crushed ice

Directions: 15 min

1. Wash and spin dry watercress; put some sheets aside for the garnish.

2. Peel the onion and cut it into small cubes. Wash the cucumber half, halve lengthways and cut the pulp into very small cubes; Set aside 4 tablespoons of cucumber cubes.
3. Puree the remaining cucumber cubes with cress, onion cubes, lemon juice, mineral water and ice in a blender.
4. Season the smoothie with salt and pepper, pour into 2 glasses and sprinkle with cucumber cubes and cress leaves.

Nutritional values | Calories 40

Protein 2 g, Carbs 5 g, Fiber 2.4 g

Cucumber and Orange Drink

Ingredients

- ½ cucumber
- 1 bunch mint
- ½ lime juice
- 3 oranges

Directions: 20 min

1. Wash the cucumber half, halve lengthways, core and dice finely.
2. Wash mint, shake dry and pluck leaves.
3. Puree the cucumber and mint in a blender.
4. Squeeze half of lime and oranges and mix with the cucumber puree.

Nutritional values | Calories 50

Protein 1 g, Carbs 9 g, Fiber 1 g

Green Smoothies with Yogurt

Ingredients

- 200 g green asparagus
- Salt
- 80 g peas
- 1 banana
- 1 tbsp. lemon juice
- 1 little apple
- 100 g baby spinach
- 1 handful apple mint (Hain mint)
- 400 g yogurt
- 100 ml mineral water or apple juice
- 1 pinch sugar
- 2 radishes

Directions: 20 min

1. Peel the asparagus in the lower third and cut off the woody ends. Cook in boiling salted water with the peas for about 8 minutes. Then pour off, quench ice-cold and let drain. Cut the asparagus tips about 8 cm long and set aside for the garnish.
2. Peel and cut the banana into pieces. Mix with the lemon juice. Peel the apple, cut it into small pieces and mix it with the banana. Wash the spinach thoroughly. Rinse off the mint and pluck the leaves. Put in the blender together with the fruit, vegetables and yogurt and mash finely. If necessary, add a little

water or juice to the desired consistency. Season with a pinch of sugar and salt.

3. Clean, wash and cut the radishes into thin slices. Halve the asparagus tips lengthways. Spread the smoothie over glasses and serve garnished with the radishes and asparagus.

Nutritional values | Calories 143

Protein 9.8 g, Carbs 27 g, Added Sugar 1 g, Fiber 4 g

CHAPTER ELEVEN

Salad and Soup Recipes

Cauliflower Soup with a Mackerel Fillet

Ingredients

- 500 g small cauliflower (1 small cauliflower)
- 1 clove of garlic
- 1 onion
- 1 tbsp. olive oil
- 1 tsp. red curry paste (or curry powder)
- 250 ml classic vegetable broth
- 250 ml milk (1.5% fat)
- Salt
- Pepper
- ½ bundle chives
- 2 stems parsley
- 80 g smoked mackerel fillet
- ½ tsp. sesame oil

Directions: 15 min

1. Clean the cauliflower, divide it into florets and wash it in a sieve. Drain well.
2. Peel and finely chop the clove of garlic and onion.

3. Heat the oil in a pot. Add the curry paste and roast for about 1 minute.
4. Stir in the garlic and onion and sauté until translucent. Add the cauliflower florets and braise everything for another 3 minutes, stirring.
5. Pour in vegetable broth and milk and simmer over medium heat for about 10 minutes. Season with salt and pepper.
6. In the meantime, wash the chives and parsley and shake dry. Cut the chives into rolls. Pluck the leaves from the parsley and chop them roughly.
7. Cut the mackerel fillet into small pieces.
8. Puree the cauliflower directly in the pot with a hand blender. Add mackerel pieces and heat briefly.
9. Season the cauliflower cream soup with sesame oil and spread over 2 bowls or deep plates. Sprinkle with chives and parsley.

Nutritional values | Calories 256

Protein 18 g, Fat 15 g, Carbs 11 g, Fiber 7 g

Arugula Cream Soup with Parmesan

Ingredients

- 150 g potatoes (2 potatoes)
- 2 shallots
- 1 clove of garlic
- 1 tbsp. rapeseed oil

- 1 l classic vegetable broth
- 100 ml cooking cream
- 240 g arugula (3 bunches)
- 50 ml buttermilk
- Salt
- Pepper
- 20 g parmesan (1 piece)

Directions: 15 min

1. Peel, wash and roughly grate potatoes on a grater. Peel the shallots and garlic and dice finely.
2. Heat the oil in a large saucepan and fry the shallot and garlic cubes until translucent. Add the potatoes and steam briefly.
3. Add broth and cream, bring to the boil and cook for 10 minutes over medium heat.
4. Wash, clean, and spin dry arugula. Put some leaves aside. Finely chop the rest with a large knife, add to the soup and bring to the boil once.
5. Add buttermilk and puree with a hand blender. Salt and pepper. Grate the parmesan finely, add to the soup with rocket leaves and serve.

Nutritional values | Calories 140

Protein 5 g, Fat 8 g, Carbs 10 g, Fiber 2.5 g

Cold Melon and Tomato Soup with Yogurt and Basil

Ingredients

- 1 organic lemon
- 1 kg small watermelon (0.5 small watermelons)
- 400 g peeled tomatoes (can; filling quantity)
- Sea-salt
- Pepper
- 4 stems basil
- 4 tbsp. yogurt (3.5% fat)

Directions: 15 min

1. Wash the lemon hot, rub dry, and finely grate half of the peel. Halve and squeeze the lemon.
2. Halve the melon with a large knife. Cut, core, peel, and dice in columns. Place in a tall container with the peeled tomatoes and liquid.
3. Add the lemon zest and juice and puree with the hand blender. Salt, pepper and cover and leave to stand in the refrigerator for 1-2 hours. Wash the basil just before serving, shake it dry and pluck the leaves. Arrange the soup with basil leaves and 1 tablespoon of yogurt each.

Nutritional values | Calories 80

Protein 2 g, Fat 1 g, Carbs 14 g, Fiber 1 g

Tomato Soup with Roasted Buckwheat

Ingredients

- 150 g onions (3 onions)
- 2nd garlic cloves
- 100 g carrots (1 carrot)
- 3 branches thyme
- 100 g celery (2 stalks)
- 175 ml classic vegetable broth
- 1 kg ripe tomatoes
- 30 g buckwheat
- 2 tbsp. olive oil
- 12 sage leaves
- Coarse salt
- Pepper

Directions: 30 min

1. Peel onions and garlic and cut into small cubes.
2. Wash, peel and dice the carrot. Wash the thyme, shake it dry and pluck the leaves.
3. Wash and clean the celery, if necessary, untangle and cut into thin slices.
4. Bring onions, garlic, carrots, celery and thyme with the broth to a boil in a saucepan over high heat. Reduce heat and cover and simmer over low heat for 10-12 minutes.

5. In the meantime, wash the tomatoes, cut out the stem ends in a wedge shape and cut the tomatoes into cubes.
6. Add to the broth and continue to cook covered for 10 minutes.
7. While the soup is cooking, roast the buckwheat in a pan without fat until golden brown and let it cool.
8. Heat the oil in a small pan and fry the sage leaves briefly in portions (attention, splashes!). Take out with a foam trowel and drain on kitchen paper.
9. Puree the soup with a hand blender and pass through a fine sieve into a second saucepan. Heat again briefly, salt and pepper. Sprinkle with buckwheat, garnish with fried sage and serve.

Nutritional values | Calories 234

Piche Steiner Stew

Ingredients
- 400 g stuck potatoes
- 500 g large carrots (5 large carrots)
- 350 g large onions (5 large onions)
- 2 garlic cloves
- 500 g savoy cabbage (0.5 head)
- 200 g lean lamb (from the leg)
- 200 g pork (from the top shell)
- 150 g lean beef goulash

- Salt
- Pepper
- 2 branches thyme
- 2 branches rosemary
- 2 stems marjoram
- 2 tbsp. rapeseed oil
- 700 ml classic vegetable broth

Directions: 2 h

1. Wash, peel and slice potatoes and carrots.
2. Peel the onions and garlic and cut them into fine slices.
3. Clean and wash savoy cabbage. Remove the stalk and cut the cabbage into broad strips.
4. Cut all types of meat into approximately 2 cm cubes and season with salt and pepper. Wash thyme, rosemary and marjoram and shake dry.
5. Heat the oil in a sealable, ovenproof pot or in a small roasting pan and sear the meat thoroughly in portions all around. Take off the stove.
6. Remove 2/3 of the meat from the saucepan and place half of the prepared vegetables on the meat in the saucepan. Season with salt and pepper.
7. Put 1/3 of the meat set aside back in the pot. Spread the remaining vegetables on top, salt and pepper.

8. Put the remaining meat in the pot and spread the herbs over it. Pour in the broth, cover and bring to a boil. Cook the stew in the preheated oven at 175 ° C (fan oven: 150 ° C, gas: speed 2) on the middle shelf for 90 minutes. Serve the Piche stein stew straight from the saucepan.

Chicken Soup the Grandmother's Way

Ingredients
- 1½ kg chicken (1 chicken)
- 3 onions
- 2 bay leaves
- 12 black peppercorns
- Salt
- 300 g celeriac (0.5 celeriac)
- 400 g large carrots (3 large carrots)
- 150 g small leek (1 small leek)
- 150 g parsnips (2 parsnips)
- 150 g parsley root (3 parsley roots)
- 200 g Hokkaido pumpkin (1 piece)
- 175 g whole grain vermicelli
- 2 stems lovage

Directions: 1 h 10 min
1. Wash the chicken, put it in a saucepan and bring to the boil, covered with 3 l of water.

2. Remove the foam rising upwards with a foam trowel.
3. In the meantime, unpeel the onions in half and roast them vigorously in a pan on the cut surfaces over high heat without fat.
4. Add onions with bay leaves, peppercorns and a little salt to the skimmed broth, simmer for 15 minutes on low heat, skimming if necessary.
5. Peel and clean half of the celery and carrots. Clean and wash half of the leek. Roughly dice everything.
6. Put the prepared vegetables in the saucepan and cook over medium heat for 1 1/2 hours.
7. Remaining celery, remaining carrots, cleaning and peeling the parsnips and parsley roots. Clean and wash the pumpkin and remaining leek. Cut everything into 2 cm cubes or slices.
8. Take the chicken out of the soup. Remove the skin and detach the meat from the bones.
9. Cut the meat into 2 cm cubes and set aside.
10. Pour the chicken soup through a sieve into a second saucepan, cook the diced vegetables in it over medium heat for 10-15 minutes. Boil the pasta in salted water, drain, hold briefly under running, cold water (frighten), then add to the chicken soup with the meat and heat. Wash lovage, shake dry and pluck the leaves. Serve the chicken soup sprinkled with the leaves.

Nutritional values | Calories 512

Protein 64 g, Fat 13 g, Carbs 32 g, Fiber 11.5 g

Caldo Verde - Portuguese Kale Soup

Ingredients

- ½ kg of potatoes or just a few medium pieces of
- about 2-3 handfuls of chopped kale (without thick stalks)
- Less than 1 liter of vegetable broth
- 1 white onion
- 1 clove of garlic
- 1 tablespoon of olive oil
- 1-2 tsp. smoked peppers (for soup and serving)
- Salt, pepper
- Toppings: smoked tempeh, fried tofu, crispy baguette (optional)

Directions: 35 min

1. Fry finely chopped onion in olive oil and grated garlic on a grater.
2. Add the previously peeled and diced potatoes and fry for about 10 minutes together with the onion.
3. Then pour the whole broth and cook until the potatoes are soft.
4. Pull out half the potatoes one after the other and set aside in a bowl for a while, and mix the soup in a pot with a hand blender. Then add the rest of the potatoes and chopped kale pieces. Cook for a few minutes until the kale softens and has a light green color.

5. Season the soup with generous pepper and salt and smoked paprika.
6. Serve with fried tempeh or tofu and eat with a crispy roll.

Avocado and Mozzarella Salad Bowl

Ingredients

- 60 g avocado
- Tomato 50 g
- Mozzarella cheese 40g
- Endive salad 20 g
- Raw lamb lettuce 10 g
- 2 tablespoons olive oil
- 1 fresh basil
- Sea salt 1 pinch
- Black pepper 1 pinch

Directions: 20 min

1. Wash the lettuce leaves, drain well, then chop them into small bowls.
2. Wash and slice the tomatoes.
3. Slice mozzarella cheese.
4. Wash the basil, shake to dry, and peel the leaves.
5. Remove the core by halving the avocado.
6. Remove the avocado pulp from the skin and cut it into strips.
7. Add avocado, tomato, mozzarella, and basil to the salad.

8. Sprinkle with olive oil and season with a teaspoon of salt and pepper in a salad bowl.

Nutritional values | Calories 336

Protein 8g, Fat 27g, Carbs 19g, Added Sugar 12 g, Fiber 5g

Chicory and Orange Salad

Ingredients

- 2 chicory
- 2 oranges
- 1 tsp. honey
- 1 tbsp. light vinegar, e.g., B. white balsamic vinegar
- 2 tablespoons of walnut oil
- Salt and pepper
- Walnut kernels as desired

Directions: 10 min

1. Wash the chicory and cut into rough strips.
2. Fillets of 1.5 oranges, squeeze out the remaining half of the orange.
3. Mix the orange juice, honey, vinegar, oil, salt, and pepper well.
4. Add chicory and orange fillets and let steep for half an hour. Sprinkle with chopped walnut kernels as desired.

Mango and Avocado Salad with Watercress

Ingredients

Salad:

- 3 handfuls of watercress (approx. 150 g)
- 1 mango
- 1 avocado
- 1 spring onion
- 2 tbsp. coriander leaves
- A little lime juice (lemon juice is of course also possible)

Dressing:

- 2 cloves of garlic, finely diced
- 2/3 tsp. ginger, finely diced
- 3 tbsp. light sesame oil
- 1 tablespoon of rice vinegar
- Salt and pepper
- Some brown sugar
- 1 chili pepper without seeds, amount depending on the spiciness and taste, very finely diced
- A little dark sesame oil

Directions 30 min

Salad:

1. Clean, wash, spin dry watercress and cut it into bite-size pieces.
2. Peel the mango and avocado, remove the stones and cut the pulp into wedges. Immediately drizzle the avocado with a little lime juice so that it doesn't turn brown.
3. Clean, wash and cut the spring onions into thin rings.

4. Pluck coriander leaves a little smaller, depending on the size.

Dressing:

1. Sweat the garlic and ginger in half a tablespoon of heated light sesame oil, place in a bowl and let cool.
2. Mix in the vinegar, salt, pepper, and sugar.
3. Fold in the remaining light oil.
4. Add the chili to taste.
5. Season with the dark sesame oil.

Nutritional values | Calories 290

Protein 5 g, Carbs 15 g, Added Sugar 8 g, Fiber 5 g

Coconut Pancakes with Kiwi Salad

Ingredients

Kiwi Salad

- 4 kiwi fruits
- 4 tbsp. maple syrup
- 2 tbsp. lemon juice

Pancakes

- 250 g flour
- 3 level teaspoons of baking powder
- 1.5 tablespoons of sugar
- Salt
- 4 eggs
- 200 ml of coconut milk

- Clarified butter

Directions: 15 min

Kiwi Salad

- Peel and dice the kiwi fruit. Then mix with the maple syrup and lemon juice.

Pancakes

- Mix the flour, baking powder, sugar and 1 pinch of salt in a bowl.
- Add the eggs and coconut milk and mix everything with the whisk to a smooth dough.
- Heat a coated pan and melt approx. 1/2 tbsp. clarified butter in it.
- Put the dough in tablespoon-sized portions in the pan and bake for 3-4 minutes on each side until golden brown.
- Process the rest of the dough as well, adding a little clarified butter if necessary.
- Serve the pancakes with the kiwi salad.

Nutritional value | Calories 482

Protein 11g, Fat 23.4 g, Carbs 58.5 g, Added Sugar 6.7 g, Fiber 8.9 g

Cucumber and Radish Salad with Feta

Ingredients

- 1½ cucumbers
- 1 bunch radish
- 1 bunch rocket (80 g)

- 4th gherkins
- 200 g feta
- 4 tbsp. olive oil
- 3 tbsp. lemon juice
- 1 tsp. mustard
- 1 tsp. honey
- Salt
- Pepper

Directions: 15 min

1. Clean, wash and cut cucumber and radishes into thin slices. Wash the rocket and shake it dry. Halve the pickled gherkins lengthways and cut into slices. Crumble the feta.
2. Whisk for the dressing oil with lemon juice, mustard, and honey, season with salt and pepper. Mix the cucumber, radish, and pickled cucumber slices and mix with the dressing. Place on a plate and sprinkle with the feta and rocket.

Nutritional values | Calories 273

Protein 10 g, Fat 23 g, Carbs 7 g, Added Sugar 1 g, Fiber 2.8 g

May Beet Salad with Cucumber

Ingredients

- 3 May turnips
- 1 cucumber
- 1 spring onion

- 2 stems parsley
- 150 g greek yogurt
- 1 tbsp. apple cider vinegar
- 1 tsp. honey
- 1 tsp. mustard
- Sea-salt
- Cayenne pepper
- Pepper

Directions: 15 min

1. Clean, peel and slice the turnips. Clean and wash the cucumber and also slicer. Clean, wash and cut the spring onions into rings. Put everything in a salad bowl and mix.
2. Wash parsley, shake dry and chop finely. Mix together a dressing with yogurt, apple cider vinegar, honey, mustard and 2–3 tbsp. water. Season with salt and cayenne pepper.
3. Mix the salad dressing with the mayonnaise and cucumber and let it steep for about 10 minutes, then grind it with pepper and serve.

Lentil Salad with Spinach, Rhubarb and Asparagus

Ingredients

- 100 g beluga lentils
- 2 tbsp. olive oil
- Salt

- 250 g white asparagus
- 100 g rhubarb
- 1 tsp. honey
- 50 g baby spinach (2 handfuls)
- 20 g pumpkin seeds

Directions: 25 min

1. Bring the beluga lentils to the boil with three times the amount of water. Cook over medium heat for about 25 minutes. Drain, rinse and drain. Mix with 1 tablespoon of olive oil and a pinch of salt. In the meantime, wash, clean, peel and cut asparagus into pieces. Wash, clean and cut the rhubarb into pieces.
2. Heat 1 tablespoon of olive oil in a pan and fry the asparagus in it for about 8 minutes over medium heat, turning occasionally. Then add rhubarb and honey and fry and salt for another 5 minutes. Wash spinach and spin dry. Roughly chop the pumpkin seeds.
3. Arrange spinach with lentils, asparagus, and rhubarb on two plates and serve sprinkled with pumpkin seeds.

Nutritional values | Calories 324

Protein 19 g, Fat 16 g, Carbs 26 g, Added Sugar 2.6 g, Fiber 13.2 g

May Beet Salad with Cucumber

Ingredients

- 3 May turnips

- 1 cucumber
- 1 spring onion
- 2 stems parsley
- 150 g greek yogurt
- 1 tbsp. apple cider vinegar
- 1 tsp. honey
- 1 tsp. mustard
- Sea-salt
- Cayenne pepper
- Pepper

Directions: 20 min

1. Clean, peel and slice the turnips. Clean and wash the cucumber and also slicer. Clean, wash and cut the spring onions into rings. Put everything in a salad bowl and mix.
2. Wash parsley, shake dry and chop finely. Mix together a dressing with yogurt, apple cider vinegar, honey, mustard and 2–3 tbsp. water. Season with salt and cayenne pepper.
3. Mix the salad dressing with the mayonnaise and cucumber and let it steep for about 10 minutes, then grind it with pepper and serve.

CONCLUSION

The Sirt Diet is to be a way to a slim figure and a long, healthy life. Scientific studies show that, indeed, sirtuins can be important for health, slowing aging and maintaining healthy body weight. Sirtuins participate in the repair of damaged genetic material, inactivation of free radicals, carbohydrate and fat metabolism, and inappropriate conditions slow down the aging process.

Sirtuins are naturally activated using a low-calorie diet, which provides 50-70 percent. Daily energy demand. These compounds also activate some polyphenols. The helpful effect of the Sirt Food Diet in slimming may result from lowering blood glucose levels, increasing insulin sensitivity of cells, as well as increasing energy production in the presence of sirtuins, as demonstrated in studies in mice.

According to the authors of the sirt diet, the effect of its use by physically active people was a decrease in body weight by an average of 3 kg per week in the first stage of the diet. Everyone declared a clear improvement in their well-being and no muscle mass loss. However, there are skeptical voices saying that the loss of such a large amount of body fat during the week is impossible, and weight loss is largely due to glycogen and water loss.

This result is attributed to a low calorie diet rather than sirtfood products alone. Products rich in polyphenols that are the basis of the sirt diet is food with very wide health values, and its positive effect on the body is confirmed by many scientific studies. Hence, it is worth including them in your diet, even if they do not have a slimming effect.

The Sirtuin Diet is full of healthy food, but not healthy eating habits. Not to mention that her theory and health claims are based on a lot of initial scientific evidence. Although adding some Sirt foods to your diet is not a bad idea and can even bring some health benefits, the diet itself looks like another fashion that will slim down your wallet more than your figure.

www.ingramcontent.com/pod-product-compliance
Lightning Source LLC
Chambersburg PA
CBHW071404210526
45465CB00001B/243